BEING AND NECESSITY

Author: A. J. Rowan

ABSTRACT

The modern disputes concerning causality and necessity—Hume's skepticism, Kant's synthetic *a priori* in answer to Hume's empiricism and skepticism, Quine's attack on analyticity, and Kripke's necessary *a posteriori*—turn on where universality and necessity are grounded: in experience, language, the categories of the intellect, or being itself. For Aristotle and Aquinas, our knowledge begins in sensation but does not end there. Through abstraction, the intellect grasps what things are and finally being as such, from which it recognizes first principles that are both universal and necessary.

This book argues that classical philosophy, drawing from Aristotle, grounds the principles of identity, non-contradiction, and causality in experience by abstraction, thereby attaining universality and necessity from particulars as features of being itself. Part I develops an answer to Hume's skepticism concerning causality. Part II reformulates Kant's synthetic *a priori* in classical philosophical terms: universality and necessity are discovered in being rather than imposed by the mind's categories. Part III extends the analysis to the analytic–synthetic debate and contemporary modal semantics, showing that the classical philosophical standpoint both anticipated and integrated insights from Frege, Quine, and Kripke arguing that the early modern epistemological turn of Hume and Kant and the turn by Frege, Quine, and Kripke in analytic philosophy were unnecessary.

The book advances three theses. First, Hume's critique of causality mistakes an epistemic description (constant conjunction) for an ontological relation (dependence of potency upon act). Second, Kant's synthetic *a priori* can be reformulated using classical philosophy: universality and necessity are attained from experience by abstraction as properties of being, not imposed by transcendental categories. Third, classical philosophy offers a metaphysical realism that dissolves the analytic–synthetic impasse and anticipates modern modal insights while addressing Quinn's attack on the synthetic *a priori* while reformulating Kripke's modal framework using Frege's logic. Reformulating Frege's formal existential quantifier to entail reference to being and experience.

The book also uses this reformulation in various proofs that justify the use of the reformulation in logic and metaphysics. It is also used to argue for classical motifs in philosophy neglected by modern philosophy, phenomenology, analytic philosophy, cosmology, biology, and various other disciplines. It is an argument for a revised view of classical philosophy and a new reformulation of the categories of necessity, causality, universality and their use in the sciences, logic, theology, philosophy of science, and metaphysics.

Table of Contents

INTRODUCTION

The dispute over causality and necessity has often been framed as a contest between habit, meaning, and mind. Hume casts necessity as custom; logical empiricists collapse it into convention; Kant secures it via the mind's categories. The classical alternative grounds necessity in being itself, accessed through experience by intellectual abstraction.

The central disputes about causality and necessity—Hume's regularity account, Kant's synthetic *a priori*, Quine's critique of analyticity, and Kripke's revival of de re modality—turn on where universality and necessity are grounded: in habits of expectation, in meanings, in the mind's categories, or in being itself. The classical view begins in sensation but reaches universals by abstraction, culminating in a grasp of being as such (ens commune). From that vantage, first principles are recognized as necessary conditions of reality, not conventions of language, nor categories of the mind, nor subjective experience and limited perception.

"Truth is the adequation of thing and intellect."[1]

On this view, the laws of thought are true because they mirror the laws of being. Identity and non-contradiction articulate the basic intelligibility of what is. Causality articulates the dependency of potency on act in the analysis of change. Because truth is the conformity of intellect to things, laws of thought express laws of being. Thus, identity and non-contradiction articulate how anything must be to be intelligible. The same method extends to causality and modal necessity.

---

1 Veritas est adaequatio rei et intellectus. — Thomas Aquinas, De Veritate, q.1, a.1 (Leonine, 1970), p. 5.

"From the memory of many instances there comes one experience; and from experience, a single universal judgment."[2]

Aristotle's sketch of induction by experience is not mere counting; it culminates in insight into universals. Aquinas radicalizes this by holding that the first known is being as such (*ens*), from which the intellect immediately sees the principles of identity and non-contradiction. We will see how this is formulated in classical modal logic and in the proposed reformulation of the synthetic *a priori*.

Thesis: universality and necessity are grounded in being and recognized by the intellect through abstraction; they are *synthetic* and *a priori* once being is grasped, but not imposed by mind.

Plan: Sections 1–3 establish the first principles; 4–6 treat Hume and Kant; 7–9 address analytic challenges and modal logic; Section 10 gathers the resolution. The remaining chapters illustrate the method and demonstrate being entails necessity.

---

2 ἐκ τῆς μνήμης πολλῶν τὸ αὐτὸ ἓν ἐμπειρία γίνεται· ἐκ δ' ἐμπειρίας ἢ καθόλου μία κρίσις. — Aristotle, Posterior Analytics II.19, 100a3–5, ed. W. D. Ross (Oxford, 1949).

1.

ARISTOTLE: BEING, IDENTITY, AND NON-CONTRADICTION

Aristotle treats the principle of non-contradiction (PNC) as the firmest principle: any denial presupposes it and all demonstration is grounded by it. Aristotle calls the principle of non-contradiction (PNC) 'the firmest' because all demonstrations presuppose it. PNC is not merely a linguistic tool or a formal logical operation; it articulates the exclusion built into being: the same cannot both be and not be in the same respect at the same time.

> "It is impossible that the same attribute should at the same time belong and not belong to the same subject in the same respect."[3]

Dialectically, meaningful assertion collapses without PNC; ontologically, being itself cannot host joint affirmation and denial under the same aspect; epistemically, PNC is *per se nota* once being is grasped. To deny PNC is performatively incoherent; ontologically, form grounds unity such that the respect in question fixes what the subject is. Classical ontology or metaphysics and epistemology inherits this from Aristotle as a law of being. Later we will discuss in detail quantum physics and the law of non-contradiction.

> "From experience the universal arises in the soul."[4]

---

3 ἀδύνατον τὸ αὐτὸ ἅμα ὑπάρχειν τε καὶ μὴ ὑπάρχειν τῷ αὐτῷ καὶ κατὰ ταὐτόν. — Aristotle, Metaphysics Γ 3, 1005b19–20, ed. W. Jaeger, OCT (Oxford, 1957).
4 ἐκ δ' ἐμπειρίας τὸ καθόλου ἐν τῇ ψυχῇ γίγνεται. — Aristotle, Posterior Analytics II.19, 100a5–6, ed. W. D. Ross (Oxford, 1949).

Thus universality is attained from particulars by abstraction. The same ascent prepares the way for causal analysis.

"Everything that is in motion is moved by something."[5]

Motion is the actuality of what exists potentially, insofar as it is potential; the mover must be actual in the relevant respect. The principle will underwrite classical causal maxims. For Aristotle on PNC and motion, the classical route distinguishes the order of discovery from the order of being: experience occasions the insight, while being attained from experience supplies the ground for the PNC, the law of identity, and the law of excluded middle.

Aristotle writes in *Posterior Analytics* II.19 (99b35–100a13) that "sense perception furnishes the universal in us." We begin with repeated experiences of particular things, which are retained in memory. From memory comes experience (*empeiria*), and from experience the intellect (*nous*) is able to grasp universals. From repeated sensory experiences of things being determines that a tree is a tree, not both a tree and not-a-tree, and the mind abstracts the universal recognition that *what is, is, and cannot at the same time not be.*

---

5 πᾶν τὸ κινούμενον ὑπό τινος κινεῖται. — Aristotle, Physics VIII.5, 257a7–8, ed. W. D. Ross (Oxford, 1936).

2.

AQUINAS: FROM EXPERIENCE TO FIRST PRINCIPLES

First Principles are **discovered from experience**, but not as a generalization. Instead, the intellect, reflecting on being through the data of sense, *sees* that contradiction is impossible. As Aristotle explains: "From perception there comes memory, and from memory (when it occurs often in connection with the same thing) experience; and from experience, universal judgments arise... so that from perception there comes to rest the universal in the soul" (*Posterior Analytics* II.19, 100a6–9). Aquinas makes Aristotle's point explicit. In *De Veritate* q.10, a.6, he explains: "The intellect knows first principles naturally, but not without the help of the senses. For nothing is in the intellect that was not first in the senses." (*nihil est in intellectu quod non prius fuerit in sensu*)."

In *Summa Theologiae* I, q.84, a.6, he develops the same idea: "The principles of demonstration are naturally known to us, since as soon as we understand the terms we know the principle. Yet this understanding comes through abstraction from sense." For Aquinas, First principles (identity, non-contradiction, causality) are *per se nota* — evident in themselves once grasped. But they are not innate. They are discovered only *through experience*, because the intellect must abstract from the data of sense. One can see what this looks like between Aristotle and Aquinas as follows:

- **Aristotle:** Experience → Memory → Experience → *Nous* grasps first principles.

- **Aquinas:** Senses → Abstraction → Intellect naturally sees first principles in light of being.

Both agree that universality and necessity are not imposed by the mind (contra Kant) but discovered in being itself, once experience provides the material. Aristotle's principle of non-contradiction is *discovered through experience* of determinate beings; Aquinas clarifies that the intellect, by abstraction from experience, naturally apprehends it as a universal necessity in light of being.

From its inception, philosophy has sought what Aristotle calls *archai*—first principles and causes. In the *Metaphysics* Aristotle remarks that "all men by nature desire to know" (980a21), and this desire finds rest only in those principles that are not themselves derivative but underlie every other act of knowing. Such principles are indemonstrable in the strict sense, not because they lack rational justification but because they provide the basis for all demonstrations. They are, as Aristotle notes, "the firmest of all" (*Metaphysics* IV, 1005b19).

Among these, the principles of identity (*what is, is*) and non-contradiction (*the same thing cannot both be and not be at the same time and in the same respect*) have long been recognized as foundational. Less obvious, but no less indispensable, is a third: the principle of causality. Scholastic tradition expressed it in two related maxims: *whatever is moved is moved by another* (*quidquid movetur, ab alio movetur*), and *nothing is the cause of itself* (*nihil est causa sui*). At its root lies the ontological claim that *what is in potency cannot, as such, reduce itself to act.*

The most primitive affirmation of thought is identity: *what is, is.* This tautological-sounding principle expresses the undeniability of being's self-sameness. Aristotle presupposes it when he writes, "for the same thing to belong and not to belong simultaneously to the same subject is impossible" (Metaphysics IV, 1011b26–27). To deny identity is to dissolve the possibility of coherent discourse, for any statement already affirms that something is what it is.

Closely allied is the principle of non-contradiction (PNC), which Aristotle calls "the firmest of all principles" (Metaphysics IV, 1005b19). Its canonical formulation is: *it is impossible for the same attribute at once to belong and not belong to the same subject in the same respect* (1005b19–20). Aquinas echoes this in *De Veritate* q.1, a.12, identifying PNC as the most certain of principles, presupposed by every act of thought.

Aquinas is just one example of many scholastics that formulate classical metaphysics and epistemology in terms of the empirical origin of knowledge. This is consistent with modern science. The classical view accepts the empirical origin of knowledge but insists that being as such is first in explanation. The intellect abstracts universal natures from particulars and recognizes first principles as necessary in virtue of being.

"Our intellect knows things by abstraction from phantasms."[6]

Abstraction explains how necessity and universality arise without innate propositions. Identity and non-contradiction are not demonstrated; they are recognized or intellectually attained once being or existence is understood. For classical epistemology, the intellect begins from phantasms or sensory data supplied by the senses and abstracts universal natures using cognitive abstract thought; yet the first object known is being, by which the intellect recognizes first principles as *per se nota*.

"To take away the cause is to take away the effect."[7]

"In efficient causes it is impossible to proceed to infinity *per se*."[8]

Likewise, by means of abstraction from sensory data, the intellect attains the principle of causality. Intellect attains the necessity and universality entailed in the ontology: to take away the cause is to take

---

6 Intellectus noster cognoscit res per abstractionem a phantasmatibus. — Thomas Aquinas, Summa Theologiae I, q.84, a.7, Leonine ed., vol. 5 (Rome, 1889), p. 318.

7 Sublata causa, tollitur effectus. — Thomas Aquinas, Summa Theologiae I, q.2, a.3, Leonine ed., vol. 4 (Rome, 1888), p. 31.

8 In causis efficientibus non est procedere in infinitum *per se*. — Thomas Aquinas, De Potentia, q.3, a.5, Leonine ed., vol. 21 (Rome, 1918), pp. 45–46.

away the effect. Similarly, that in efficient causes it's impossible to proceed to infinity *per se*. It is necessary for a chain of causality to terminate. These are not probabilistic summaries nor mathematical models, but articulations of *per se* causal order: derivative causes act only so long as a principal cause confers actuality here and now. A complete list of first principles follows comparing the classical list of first principles to contemporary first principles.

Domain	Classical Formulation	Contemporary Expression	Status Today
Identity	*What is, is*	$\forall x \, (x = x)$	Universally accepted
Non-Contradiction	*Nothing both is and is not*	$\neg(P \land \neg P)$	Universally accepted (though paraconsistent logics explore exceptions)
Excluded Middle	*Everything either is or is not*	$P \lor \neg P$	Accepted in classical logic; rejected in intuitionistic logics
Bivalence	—	Every proposition is true or false	Contested (vagueness, future contingents, quantum indeterminacy)
Causality	*Potency requires act*	Causal dependence, counterfactuals, grounding	Widely debated (regularity vs. necessitarian vs. dispositionalist accounts)
Finality	*Every agent acts for an end*	Teleology / directedness	Controversial; revived by some neo-Aristotelians
Sufficient Reason	*Nothing is without reason*	Every fact has an explanation	Central in rationalist metaphysics; disputed
Grounding	—	Facts grounded in more fundamental facts	Important in contemporary metaphysics

In summary, for classical metaphysics abstraction and the *per se* causal order are attained from being: experience occasions the insight, while being or existence itself supplies the ground. This is not induction in the modern sense. The mind does not add up many cases and conclude probabilistically. Instead, experience provides the *occasions* for the intellect to **see** something necessary about being. For example: after many perceptions of determinate beings, the intellect sees that "what is, is" and "it cannot be both A and not-A in the same respect." Thus, for classical metaphysics and epistemology, **experience occasions, but does not generate, first principles.** They are grasped by *nous* (intellect), not by induction. Note that Aquinas like many medieval philosophers of mind attempt to ground this process in the process of visual or sense perception very similar to the way a physiologist might or in computer vision.

3.

CAUSALITY AS PARALLEL TO IDENTITY AND NON-CONTRADICTION

Causality functions as a first principle parallel to identity and non-contradiction. Identity: what is, is. Aristotle asserts: "For the same thing to belong and not to belong simultaneously to the same subject is impossible."[9] Although this text doesn't use the formula "what is, is," Aristotle presupposes the law of identity as the basis of PNC. Non-contradiction: nothing can both be and not be in the same respect. Aristotle argues that "It is impossible for the same attribute at once to belong and not belong to the same subject and in the same respect." *Metaphysics* Γ (Book IV), 1005b19–20). Then in 1006a11–12, Aristotle argues again: "The most certain of all principles is that contradictory statements are not at the same time true."

Aristotle describes causality in terms of potency and act, or causality is simply the ***motion of change from what is potential to what becomes actual***, in his work on physics, he states "Motion is the actuality of what exists potentially, in so far as it exists potentially" (*Physics* III, 201a10–12). Again, "Whatever is moved is moved by another." (*Quidquid movetur, ab alio movetur.*) (*Physics* VII).[10]

---

9 ἀδύνατον τὸ αὐτὸ ἅμα ὑπάρχειν τε καὶ μὴ ὑπάρχειν τῷ αὐτῷ καὶ κατὰ ταὐτόν. Aristotle, Metaphysics IV (Γ), ch.3, 1005b19–20.

10 Quidquid movetur ab alio movetur. — Aquinas' rendering of Aristotle, cited e.g. ST I, q.105, a.5 (Leonine).

Aristotle in his work on metaphysics argues that "Potency is a principle of change in another thing or in the thing itself qua other" (*Metaphysics* IX, 1049b4–5). It follows, that "There must be a first unmoved mover, eternal and immovable" (*Metaphysics* XII, 1072a20–26). Because causality is a movement between what potentially exists to what comes into existence, for causality: what is in potency cannot, as such, reduce itself to act. In other words, what does not yet exist cannot cause itself to exist. For every effect to exist it must necessarily have a cause for that effect to exist. And ultimately, for anything to exist at all, there must necessarily be a cause that did not move from non-existence to existence, from potentiality to actuality. This we call the Prime Mover, or First Cause.

Identity ('what is, is') and non-contradiction ('not both A and not-A in the same respect') are mirrored by causality: 'what is in potency cannot reduce itself to act.' The above classical maxims give this ontological shape. As with PNC, the necessity here is seen in light of being, not 'read off' any single observation. Hence causality is synthetic and necessary once act and potency are grasped from experience.

Classical metaphysics correctly articulated the necessity of causality in two related axioms: *whatever is moved is moved by another* (*quidquid movetur ab alio movetur*) and *nothing is the cause of itself* (*nihil est causa sui*). Its ontological formulation is: *what is in potency cannot, as such, reduce itself to act.* Potency, by definition, lacks actuality; it is the capacity to be. But capacity cannot activate itself. Actuality is required to bring potency into being. Thus, every transition from potency to act requires a cause.

This principle is indeed attained inductively from regularities in nature. Just as the PNC is attained from observations of things not being contradictory, so too causality is attained by virtue of repeated sequences of events. But unlike Hume's position, this causality exists ontologically by virtue of the way laws of nature work. Being is both an abstract concept describing all of existence, both material and immaterial, but it is also the ontological existence of everything that is. In the *Physics* (III–VIII),

Aristotle defines motion as the actuality of a being in potency, insofar as it is in potency (201a10–12). Motion is intelligible only if potency is real and requires act for its realization. Hence his conclusion: whatever is moved is moved by another. In *Metaphysics* IX, Aristotle develops act and potency as co-principles of being. Every change presupposes the priority of act, for potency as potency cannot actualize itself. Ultimately, this reasoning leads to the Unmoved Mover (*Metaphysics* XII): a first actuality that causes without itself being caused. The necessity of causality is thus the foundation for Aristotle's *Prime Mover*. What emerges from the texts is that potency does not reduce itself to act, change requires something already actual to actualize potency, and therefore, causality is a necessary principle of being itself. For extensive justification and defense of these principles see *The First Cause*.

Those like Aquinas takes Aristotle's insight and renders it into a classical metaphysical framework. In the *Summa Theologiae* I, q.2, a.3, his "second way" argues that there must be a first efficient cause, since nothing can be the cause of itself, and an infinite regress of causes would eliminate causality altogether. Aquinas argues: "In causis efficientibus non invenitur neque est possibile quod aliquid sit causa sui ipsius; quia sic esset prius seipso, quod est impossibile." The English translation is: "In efficient causes it is not possible to go on to infinity, for in all efficient causes following in order, the first is the cause of the intermediate cause, and the intermediate is the cause of the ultimate cause, whether the intermediate cause be several, or one only. Now to take away the cause is to take away the effect. Therefore, if there be no first cause among efficient causes, there will be no ultimate, nor any intermediate cause. But if in efficient causes it is possible to go on to infinity, there will be no first efficient cause, neither will there be an ultimate effect, nor any intermediate cause; all of which is plainly false. Therefore it is necessary to admit a first efficient cause, to which everyone gives the name of God."

In *De Potentia* q.3, a.5, Aquinas clarifies this by explaining that nothing can pass from potency to act except through something already in act. Potency cannot move itself; it requires act. It follows from this that no cause can be the cause of itself: "Nothing is the cause of itself."[11] Potency cannot move itself; it requires act. Every cause is reduced to act by something already in act. For this reason, no natural cause can be its own act of existence, nor can it be the cause of its own existence. This is not a contingent feature of beings but a structural truth of existence or Being itself. Aquinas writes, "Nihil potest seipsum reducere de potentia in actum, quia sic esset actu et potentia simul secundum idem."[12] The English translation, "Nothing can reduce itself from potency to act, for thus it would be actual and potential at the same time in the same respect."[13]

Thus, causality is *per se nota*—not in the sense of being analytically contained in concepts, but in the sense of being evident once act and potency are understood. Just as identity and PNC are necessary and universal, so too is causality. To deny PNC is to fall into contradiction; to deny causality is to collapse the distinction between potency and act, leaving change unintelligible. Both denials undermine discourse and knowledge.

Here lies the error of Hume. By defining causality as constant conjunction, he reduced it to an epistemic description, mistaking the shadow for the substance. The real ground of causality is ontological such that potency requires act, and so causality is necessary. Kant's attempt to save necessity through the synthetic *a priori* recognized the insufficiency of induction but relocated the source of necessity in the categories of the mind rather than in being. Aquinas offers a corrective such that necessity and universality are discovered in being, not imposed by the intellect.

Causality stands alongside identity and non-contradiction as a first principle of being. Each articulates an aspect of the structure of Being itself (all of existence) without which nothing is intelligible. Three ontological fundamental principles: identity secures that beings are what they are, non-contradiction secures that they are not what they are not,

---

11 Nihil est causa sui. — Thomas Aquinas, In Physicorum VIII, lect. 5; cf. ST I, q.2, a.3 (Leonine).
12 Thomas Aquinas, *Quaestiones Disputatae de Potentia Dei*, q.3, a.5, in *Opera Omnia*, vol. 2 (Rome: Leonine Commission, 1965), 49.
13 Thomas Aquinas, *On the Power of God (Quaestiones Disputatae de Potentia Dei)*, trans. English Dominican Fathers (Westminster, MD: The Newman Press, 1952), q.3, a.5.

causality secures that potency cannot actualize itself but depends upon act and therefore every effect has a cause by necessity.

Together these principles ground classical metaphysics. They are not hypotheses drawn from how one feels or believes experience is, but ontological necessities grasped through abstraction. Thus causality, far from being a mere habit of mind, is a universal and necessary principle that renders motion, change, and existence itself intelligible.

4.

HUME'S CRITIQUE OF CAUSALITY AND A CLASSICAL RESPONSE

In the eighteenth century, David Hume pressed what became one of the most influential critiques of causality. For Hume, all knowledge derives from impressions and ideas. The impressions are immediate sense perceptions, and ideas are faint copies of these. But when we investigate causality, we find no sensible impression corresponding to "necessary connection."

As Hume writes in the *Enquiry Concerning Human Understanding*: "When we look about us towards external objects, we are never able, in a single instance, to discover any power or necessary connection."[14] We may see the movement of one billiard ball striking another, but we never perceive the "glue" that binds the events together. The most we find is that certain kinds of events are constantly conjoined: fire follows upon flame, rain upon clouds, the sun rises each morning. From this, Hume argues, we acquire the expectation that the future will resemble the past. But expectation is a psychological habit, not a rational insight into necessity. Thus he concludes: "Custom... is the great guide of human life."[15]

---

14 "...never able, in a single instance, to discover any power or necessary connection..." — David Hume, An Enquiry Concerning Human Understanding, §7, ed. T. L. Beauchamp (Oxford: OUP, 1999), pp. 62–65.
15 "Custom... is the great guide of human life." — Hume, Enquiry, §5, same ed., p. 43.

For Hume, then, the necessity we associate with causal relations is not in the objects themselves but in our minds, trained by experience to anticipate regularities. Causality, stripped of ontological weight, becomes a projection of subjective habit onto the sequence of impressions.

The radical implication of Hume's position is that causality lacks objective necessity. It is contingent association, and all claims of necessary connection in nature are fictions of the imagination. This deeply unsettled philosophy and science. If causality is only custom, then science has no rational warrant for universal laws; it rests only on expectation that past regularities will continue.

A classical reformulation agrees that necessity is not a sensible quality but adds that intellect discerns ontological dependence: potency requires act; instruments require a principal cause; *per se* series cannot be infinite. Regularity is a symptom of dispositional structures rooted in natures.

5.

KANT AND THE SYNTHETIC
A PRIORI

Immanuel Kant famously confessed that it was Hume who "awakened him from his dogmatic slumber." Kant saw clearly that Hume was right on one point: necessity is not a sensible quality. We never observe it in the way we observe color or motion. If necessity is not given in experience, then it must come from another source.

Kant therefore distinguished between what comes *from* experience and what arises *through* experience. In the *Critique of Pure Reason* he writes: "That all our cognition begins with experience there can be no doubt; but... it does not on that account all arise from experience."[16] From experience, we learn that things are a certain way. But this does not tell us that they could not be otherwise. As Kant puts it: "Experience teaches us that something is so and so; but not that it cannot be otherwise... valid absolutely *a priori*."[17]

For Kant, the mind supplies the framework that makes experience possible. Space and time are forms of intuition, and the categories of the understanding (such as causality, substance, and quantity) are the rules by which the manifold of sense is synthesized into coherent experience. Without these categories, we would have a chaos of impressions but no unified objects of knowledge. Causality, in particular, is secured as

---

16 Daß alle unsere Erkenntnis mit der Erfahrung anfange... entspringt sie darum nicht eben alle aus der Erfahrung. — Immanuel Kant, Kritik der reinen Vernunft, B1–2 (AA III), Berlin: Reimer, 1900.
17 Die Erfahrung lehrt zwar, daß etwas so beschaffen sei, aber nicht, daß es nicht anders sein könne. Immanuel Kant, Kritik der reinen Vernunft (AA III), A92/B124, p. 63; trans. Norman Kemp Smith.

a category of the understanding. Kant insists: "All alterations occur in accordance with the law of the connection of cause and effect."[18]

This law is not derived from experience, since experience gives only conjunctions. Rather, it is the condition that makes experience of temporal succession possible at all. Unless events are ordered by the rule of causality, we could not distinguish between mere succession and objective sequence.

Thus Kant preserves universality and necessity by grounding them not in external objects but in the mind's contribution to appearances. Causality is a *synthetic a priori* judgment: it adds to our concept of events a relation of necessity, not derivable from experience but required for experience to be intelligible.

Yet this solution comes with a cost. Necessity is no longer in things themselves but only in phenomena as constituted by the subject. The noumenal realm—things in themselves—remains inaccessible.

Kant secures universality and necessity by locating them in the mind's contribution to appearances. The classical view retains the ampliation and necessity but grounds them in being rather than in forms of understanding or a transcendental synthesis. The following chapters propose a reformulation of Kant's synthetic *a priori*.

---

18 Alle Veränderungen geschehen nach dem Gesetze der Verknüpfung der Ursachen und Wirkungen. KrV A189/B232 (AA III), p. 211.

6.
A REFORMULATION OF THE SYNTHETIC *A PRIORI*

A classical approach agrees with Hume and Kant at one point: necessity is not a sensible quality. One does not see necessity in the color red or in the impact of two objects. But unlike Hume, the classical tradition does not reduce causality to custom. And unlike Kant, it does not confine necessity to the structures of the mind.

Instead, it locates necessity in being itself, once act and potency are understood. Aristotle defines motion as "the actuality of what exists potentially, insofar as it exists potentially" (*Physics* III, 201a10–12). From this he concludes, "whatever is moved is moved by another" (*Physics* VII, 241b24–26). The principle is simple: what is in potency cannot, as such, reduce itself to act. Potency requires act. Aquinas sharpens this in *De Potentia* q.3, a.5: "Nothing can reduce itself from potency to act, for thus it would be actual and potential at the same time in the same respect." And in the *Summa Theologiae* I, q.2, a.3: "In efficient causes it is not possible to go on to infinity... therefore it is necessary to admit a first efficient cause, to which everyone gives the name of God."

These maxims are not inductive generalizations. They are ontological insights grasped by the intellect once it has abstracted the concepts of act and potency from experience. The principle of non-contradiction is understood by examining being itself, not by counting examples; similarly, causality is grasped through the metaphysical concepts of potency and act. Meaning that by seeing the structure of potency and act throughout nature, concepts (*quidditates*) are formed of act and potency and their relation to one another.[19]

For Aristotle and Aquinas as examples of the classical approach, the intellect does not impose necessity upon experience, but discovers it by abstracting from the structure of being. By observing change and motion in the sensible world, the mind comes to recognize the metaphysical distinction between potency (*dynamis*) and act (*energeia* / *actus*). Potency is the capacity for being, act is the fulfillment of that capacity.

Once this distinction is grasped, the intellect forms universal concepts (*quidditates*) of act and potency, and of their relation to one another. These concepts are not empty constructs but reflections of real metaphysical principles that structure every being. Thus the intellect discerns, for example:

- that what is in potency cannot actualize itself (causality),
- that what is in act is prior to potency (priority of actuality),
- and that every finite act is received in a potency (composition of essence and existence).

In this way, the **laws of identity, non-contradiction, excluded middle, and causality** are not abstractions from linguistic usage, nor arbitrary mental categories, but **intellectual apprehensions of being itself as revealed through act and potency.**

Regularity, therefore, is not the cause of our idea of necessity, as Hume thought, but a symptom of deeper dispositional structures rooted in natures. Fire heats because its nature is such that it produces heat, not because we have seen it heat repeatedly. Stones fall because natural laws of gravity, when unimpeded, tends to the earth. The order of nature is not imposed by custom but flows from intrinsic principles of being.

The strength of the classical position is that it integrates what Hume and Kant each saw in part. From Hume it accepts that necessity

---

19 Quidditas comes *quid* ("what"), with the suffix -*itas* ("-ness" or "-ity"), literally meaning **"whatness"** or **essence. quidditas** = "whatness," the essence of a thing, *that by which something is what it is.* The term **essentia** interchangeably with *quidditas*.

is not an impression. From Kant it accepts that necessity is universal and not derived from induction. But it avoids Hume's reductionism and Kant's subjectivism by grounding necessity in the real distinction between potency and act.

The difference between Kant's and the classical approach can be summarized as follows:

- For Kant, the universality and necessity of causality belong to the subject's way of constituting experience.
- For Aristotle and Aquinas, they belong to being itself, discerned by the intellect.

This distinction has far-reaching implications. On Kant's model, metaphysics becomes impossible beyond the bounds of experience, since necessity belongs only to the forms of appearance. On the classical model, metaphysics remains possible, since causality is rooted in being, not in subjective synthesis.

Hence Aquinas can reason to a first cause: if nothing reduces itself from potency to act, and if no causal series can regress infinitely, then there must be a first actuality, pure act, that grounds all others. For Kant, such reasoning would be invalid, since causality applies only within experience. For Hume, it would be meaningless, since causality is nothing more than expectation.

Hume's skepticism revealed that necessity is not a sense-datum; Kant's transcendental philosophy preserved necessity by relocating it in the subject. But both missed the classical insight that necessity belongs to being as such. The intellect, through abstraction, grasps first principles in light of being, not as empirical habits nor as mental impositions.

Thus a classical reformulation preserves what is true in both Hume and Kant while avoiding their limitations. Causality is not a fiction of the imagination nor merely a form of thought but a first principle parallel to identity and non-contradiction: *what is in potency cannot, as such, reduce itself to act*. Aquinas as an example, the intellect discovers universality and necessity not by induction alone but by abstracting the *quidditas*—the "whatness" or essence—of things from sensible experience. As he explains in *De Ente et Essentia*, "*quiditas autem, quam significat definitio, est idem quod essentia*" ("the quidditas, which the definition signifies, is the same as the essence," ch.1). Essence or *quidditas* is that by which a thing is what it is, signified by its definition: "*essentia vero, sive quidditas, est id quod significatur per definitionem*" (ibid.). Through this process of abstraction, the intellect forms universal concepts of act and potency, since every sensible being presents itself as a composite of what it is in act and what it may be in potency.

The principle is not imposed by the mind but grasped in light of being itself. Thus, in the *Summa Theologiae* I, q.3, a.3, Aquinas can say of God alone: "*Deus est sua quidditas, seu sua essentia*" ("God is His own quidditas, or His own essence"), underscoring that in creatures, essence and existence are distinct, whereas in God they are identical. As the intellect reflects on the act–potency structure throughout nature, it forms *quidditates* of these relations, discerning that potency cannot actualize itself, that every act is prior to potency, and that finite acts are always received in a limiting essence. In this way, *quidditas* becomes the bridge between sensible experience and metaphysical necessity.

In classical metaphysics and epistemology, synthetic *a priori* truths are ampliative yet necessary from experience. First principles are grasped: identity, non-contradiction, the excluded middle, and causality through a process. They are not analytic by meaning alone, but they become evident *a priori* after the intellect understands being and the act/potency structure of the nature of things grasping that they cannot be otherwise. Hence, causal principles are synthetic *a priori*: discovered from experience via abstraction, grounded in being, not legislated by the understanding.

The first book in this series explored Kant's doctrine of the synthetic *a priori* in greater detail, the reasons why he introduced it, and the manner in which a classical metaphysical framework can reformulate it. For the fuller foundation that supports this reformulation, see the first book in this series, *The First Cause*. It underpins the modal arguments given throughout *Being and Necessity*. What follows are the details of how that metaphysical rebirth takes place, and how the new *Copernican Revolution* is formulated in a new multi-modal framework based on classical metaphysics and logic.

7.

QUINE AND KRIPKE: MODERN ANALYTIC CHALLENGES

In the mid-twentieth century, W. V. O. Quine cast doubt on one of the central assumptions of logical positivism: the distinction between analytic and synthetic truths. For the positivists, analytic truths were grounded in linguistic convention ("All bachelors are unmarried"), while synthetic truths were grounded in empirical fact. This distinction was meant to secure both logical certainty and empirical science.

But in his influential essay *Two Dogmas of Empiricism*, Quine famously declared:

> "Any statement can be held true come what may... no statement is immune to revision."[20]

For Quine, meaning is not insulated from experience but part of a holistic web of belief. No single statement is immune from revision, since any truth, even one thought analytic, can in principle be adjusted if the network of beliefs demands it. By collapsing the analytic–synthetic distinction, Quine dissolved the idea that there are truths grounded merely in convention, independent of empirical reality.

This was devastating for the Kantian project of securing synthetic *a priori* judgments. If necessity is merely conventional, then it is hostage to linguistic revision. If necessity is empirical, then it lacks absolute force. For Quine, necessity as traditionally conceived is suspect.

---

20 "Any statement can be held true come what may... no statement is immune to revision." — W. V. O. Quine, "Two Dogmas of Empiricism," Philosophical Review 60 (1951): 40.

Against this background, Saul Kripke's *Naming and Necessity* (1972) was revolutionary. Kripke argued that necessity cannot be reduced to linguistic convention or logical tautology. Instead, there are metaphysical necessities discoverable *a posteriori*. His key tool is the notion of rigid designation: "A rigid designator designates the same object in all possible worlds in which that object exists."[21]

Proper names, like "Aristotle," designate the same individual across possible worlds, not by description but by direct reference. Thus, even if Aristotle had not been a philosopher, he would still have been Aristotle.

This leads to the startling thesis of the **necessary *a posteriori***: there are truths that are necessary but knowable only through experience. For instance, "Water is H_2O" is necessary because in all possible worlds where water exists, it is H_2O. But this necessity is discovered empirically, not analytically.

In this way, Kripke rehabilitated metaphysical necessity, relocating it from linguistic convention to the structure of reality itself. Unlike Quine, who relativized necessity to revisable linguistic frameworks, Kripke insisted that necessity belongs to things as they are.

The classical tradition integrates the insights of both Quine and Kripke while avoiding their pitfalls. From Quine, it accepts the critique of analyticity as mere convention. The truths of necessity cannot be reduced to arbitrary stipulations of language. From Kripke, it accepts the recovery of metaphysical necessity through rigid designation and the necessary *a posteriori*.

But classical philosophy grounds necessity more deeply—in the natures of things. For Aristotle and Aquinas, necessity is not imposed by language nor constructed by thought. It is discovered in being itself. Potency requires act, essence grounds operation, and natures structure dispositional regularities.

The classical reformulation interprets the necessary *a posteriori* in terms of the *synthetic a priori*. The intellect, through experience, grasps truths that are not analytic by convention but necessary in light of

---

21 Saul A. Kripke, *Naming and Necessity* (Cambridge, MA: Harvard University Press, 1980), p. 97.

natures. For example, once the essence of water is discovered to be H_2O, it is recognized as necessarily so. This necessity is metaphysical, not linguistic, and the intellect apprehends it through abstraction from empirical discovery.

This approach also reflects how the applied sciences work. Physics and chemistry do not merely track appearances; they uncover the natures of things. When scientists discover the molecular structure of water, they are not revising a convention but revealing a truth about being that holds universally and necessarily. Regularities are symptoms of dispositional structures grounded in natures.

Quine undermines analyticity as linguistic convention; Kripke rehabilitates *de re* necessity and the necessary *a posteriori*. Kripke re-centers metaphysical necessity via rigid designation and the necessary *a posteriori*. A classical reformulation brings these strands together: it locates necessity in natures while acknowledging that many such necessities are discovered *a posteriori*, in the form of what Aristotle would call knowledge gained "from experience" but abstracted to universality by intellect making them synthetic *a priori*.

The classical reformulation integrates both insights by locating necessity in natures while allowing *a posteriori* discovery in the form of the synthetic *a priori*. And this is the way the applied sciences work, they discover the nature of things not merely the appearance of things. Thus, where Quine relativized necessity to convention and Kripke grounded it in metaphysical reference, classical realism grounds it in being itself. Necessity belongs neither to linguistic stipulation nor to purely mental categories but to the very structure of reality, discoverable through the sciences and intelligible in philosophy. In this way, classical metaphysics offers a framework that both absorbs and transcends the analytic challenges, showing that causality, identity, and necessity remain first principles of being.

8.

THE ANALYTIC–SYNTHETIC DIVIDE: ORIGINS, CRISIS, AND A CLASSICAL REFORMULATION

Kant introduces two orthogonal distinctions—analytic vs. synthetic and *a priori* vs. *a posteriori*—yielding a fourfold grid: analytic *a priori* (logic, definitions), synthetic *a posteriori* (empirical science), analytic *a posteriori* (idle/rare), and synthetic *a priori* (the crucial class for mathematics and the analogies of experience).

Analytic judgments are true by virtue of meanings; their denial yields contradiction. Synthetic judgments add content not contained in the subject concept. *A priori* indicates necessity and strict universality; *a posteriori* indicates empirical origin and contingency. The pressure point is the **synthetic a priori**: ampliative yet necessary.

Kant's introduction of the analytic/synthetic divide intersects four ideas: (1) the analysis of concepts (analytic), (2) ampliation of knowledge (synthetic), (3) the *a priori* mark of universality and necessity, and (4) the *a posteriori* mark of contingency and experience. The decisive novelty is not the analytic/synthetic distinction alone, but its crossing with the *a priori/a posteriori*, yielding the problematic category of the synthetic *a priori*. Kant writes:

> "In analytic judgments the predicate belongs to the subject as contained within it; in synthetic judgments the predicate lies entirely outside the concept."[22]

---

22 Analytische Urteile …; synthetische Urteile … — Kant, KrV A6–7/B10–11 (AA III).

This classification creates the puzzle: how can a judgment be both ampliative (synthetic) and yet necessary and universal (*a priori*)? If necessity cannot be read off experience, and if ampliation cannot be secured by mere analysis of meanings, what grounds synthetic *a priori* knowledge?

Two problems ensue. First, if necessity and universality cannot be derived from induction, from where do they arise? Kant's answer: the mind's pure forms and categories that structure all possible experience. Second, if synthetic *a priori* judgments apply only to appearances, can metaphysics speak about things in themselves? Kant denies this, disallowing classical metaphysics.

Kant's fourfold grid set the stage for the twentieth century. Logical empiricism depended on a sharp analytic/synthetic boundary; Quine's critique eroded it. But Analytic philosophy initially embraced the divide: Frege sharpened the analytic ideal through logicism, Russell and Whitehead pursued reduction of arithmetic to logic, and the Vienna Circle instrumentalized the divide to separate tautologies from empirically meaningful statements. The cost was that necessity seemed either a matter of linguistic convention or a product of verification—until Quine pressed the decisive objections.

Quine's two famous theses cast doubt on both pillars:

"Our statements about the external world face the tribunal of sense experience not individually but only as a corporate body... No statement is immune to revision."[23]

First, analyticity lacks a non-circular foundation; second, confirmation is holistic rather than sentence-by-sentence. The upshot was to unsettle the analytic/synthetic boundary that the logical empiricists needed. Necessity could not be quarantined as mere linguistic truth.

Kripke showed that necessity and the *a priori* come apart: some truths are necessary but known *a posteriori* (e.g., 'Water is H2O'), while some *a priori* statements are not necessary *de re*. Rigid designation explained

---

23 Our statements about the external world face the tribunal of sense experience not individually but only as a corporate body. — W. V. O. Quine, "Two Dogmas of Empiricism," *Philosophical Review* 60 (1951): 39–40.

how terms can latch onto the same entity across counterfactual scenarios, yielding necessary identity and essentialist claims.

"A rigid designator designates the same object in all possible worlds in which that object exists."[24]

This severed long-assumed alignments and re-centered metaphysical necessity. But if necessity is a feature of things, we still need an account of how the intellect comes to know it without appealing to stipulative meanings or merely to the architecture of the mind.

A classical philosophy reformulation of the synthetic *a priori* resolves the tensions: the intellect begins in experience but abstracts universality and necessity from being itself. Identity and non-contradiction are necessary because being excludes non-being. Causality is necessary because potency requires act. These are synthetic (ampliative) yet *a priori* once being is grasped, not as impositions of cognition but as discoveries about reality.

In classical philosopphy, synthetic *a priori* truths would be ampliative because they concern being (not mere meanings), and *a priori* because, once the nature is understood, the intellect recognizes that it cannot be otherwise. The route of discovery can be empirical, but the ground of necessity is *de re*. Identity and non-contradiction follow from the intelligibility of being; causality follows from the priority of act to potency. Act and potency is itself derived from experience, which is in potency can be acted upon. An apple tree is acted upon to become an apple tree. It does not simply become an apple tree without being acted upon by principles of causality acting upon a seed until it becomes the apple tree. It is the very nature or essential nature of an apple tree seed to become an apple tree and not a fig tree. The intellect abstracts the change from potentiality to act, from what it's essential nature determines for it to be to what it becomes consistent upon its nature or essentiality. It's essence or being (ens/esse) determines what the apple seed will become in moving from potency to actuality.

---

24 A rigid designator designates the same object in all possible worlds in which that object exists. — Saul A. Kripke, Naming and Necessity (Cambridge, MA: Harvard University Press, 1980), 48.

From this route of discovery one empirically from experience attains the ground of necessity is *de re*. One grasps some x must necessarily be x and not y. This principle of identity, that an apple seed must necessarily become an apple tree consistent with its essential nature leads one to grasp the first principle of identity, something cannot not be what it is. If an apple seed exists, it is an apple seed. If it's nature is such that it will become an apple tree, it will not become a fig tree. The principle of identity follows from experience. There is nothing in the intellect not first in the sense.

Likewise, the law of non-contradiction or the first principle of contradiction follows, some x cannot both be x and not x at the same time and in the same respect. An apple seed is an apple seed, that's what it is by its essential nature and cannot be otherwise, therefore it cannot both be and not be an apple seed at the same time and in the same respect. It is potentially an apple tree, not potentially a fig tree. Once acted upon, the essential nature of an apple tree necessarily determines that the apple seed will become the apple tree once the apple seed is acted upon. Thus the first principles of identity, non-contradiction, and necessity all follow from empirical senses not the categories of the mind. They are abstracted from nature itself. One sees this is the case for an apple tree, an orange tree, a child becoming an adult etc. and the intellect grasps the universality and necessity of these first principles.

Again, synthetic *a priori* truths are ampliative because they concern being (not mere meanings), and *a priori* because, once the nature is understood, the intellect recognizes that it cannot be otherwise. The route of discovery of first principles is empirical, but the ground of necessity is *de re*. That necessity is in being itself. It is in the nature of things as they exist and as they are essentially. Identity and non-contradiction follow from the intelligibility of being; causality follows from the priority of act to potency. And likewise, causality. Aquinas's principle is terse:

"To take away the cause is to take away the effect."[25]

---

25 Sublata causa, tollitur effectus. — Thomas Aquinas, Summa Theologiae I, q.2, a.3, Leonine ed., vol. 4 (Rome, 1888), 31.

The necessity at stake is not psychological habit nor belief nor feeling, but ontological structure and ontological necessity. If one removes a cause, one takes away the effect produced by that cause. Let's take heating liquid water at 212 Fahrenheit or 100 Celsius as an example. For heat to produce water vapor from liquid water a fundamental change in the motion of H20 molecules takes place and the H20 molecules move more freely. The physical state changes not the molecular structure of the water molecules. The molecules move faster and spread apart due to the transference of heat energy in conduction, but the molecular structure remains H20. Conduction transfers heat energy and conduction is the process of transferring heat energy by direct contact with molecules.

Fire itself is a result of combustion and needs three things fuel, oxygen, and a heat source such as a spark, friction, open flame, or concentrated sunlight and when combined it releases fire's energy. One can already see a causal change. If one removes any one of the causes (the oxygen, fuel, heat are causes in the fire triangle), the effect of producing water vapor is lost due to the loss of energy transfer.

Combustion transfers energy because it's a chemical reaction that releases heat and light making molecules move faster. H20 molecules remain the same but at boiling point at 212 Fahrenheit or 100 Celsius, the molecules break free from their liquid state and move apart thus becoming liquid gas. This is called thermal expansion. The heat energy from the heat causes the change from potency to act, i.e., from liquid water to vapor water. The change is a physical state change in H_2O, but NOT a molecular change as in combustion. Thermal expansion is where the heat energy causes molecules to expand and move apart, leading to expansion. When water reaches its boiling point, thermal expansion causes water molecules to move apart, turning into water vapor. If one takes away the cause in this process of thermal expansion, one takes away the effect of water vapor. Causality is not a category of the mind of a knowing subject, but a category of being itself from which the intellect apprehends this causality from being itself and the causal relationships that exist in the world. We abstract such principles from being and they become universal and necessary by virtue of being itself as the mind attains these principles of nature.

Once one understands what potency and act are, the denial of causality undermines intelligibility just as denial of non-contradiction does. Thus, when Hume denied the knowability of causality, he undermined the intelligibility of causality. This awakened Kant to argue for categories of the mind in his Transcendental philosophy to justify the knowability of causality. Kant denied traditional metaphysics and causality to provide a more secure foundation for science and empirical knowledge due to Hume undermining the intelligility of causality in nature. Hume did not understand causality nor the basis for coming to understand causality from being itself. Thus he believed it was merely an inference inferring causality from one event following another; he called it a habit of association based on our experiences of events following one from another. It is true that causality is a universal and necessary inference from repeated experiences of causal events following one from another, but this does not undermine their existence in nature nor does it make the inference unreliable nor unintelligible. Causality is not merely a habit of the mind, but it's the nature of being itself. The mind grasps in causality its universality and necessity. In other words, we attain the principle of causality from existence and attain its universality and necessity from nature itself; from being.

Contrary to Kant's transcendental philosophy, causality is attained from being itself, not from the categories of the mind. For some x to act upon y in nature *de re* is not merely a Kantian empirical perception of appearances, nor an apperception—self-awareness that synthesizes all thoughts and perceptions through the categories into a unified self-consciousness that interprets sensory data and experience—but rather a grasp of how nature itself operates.

It is neither a belief nor a habitual inference about nature, nor a universal and necessary category imposed upon it; rather, it is an essential quality and feature of nature itself—intelligible and coherent—attained through abstraction from repeated instances of direct experience. This is why H_2O changes state from liquid to gas in a universal and necessary causal expansion of its molecules, triggered by combustion arising from the union of oxygen, heat, and fuel. And this is why it remains intelligible regardless of the knowing subject.

For Kant the 12 categories are innate structures of the mind and Kant justified this by claiming they are necessary conditions for us to understand our experience and sense data. In the *Transcendental Deduction* Kant argued that the categories are necessary for coherent experiences. He believed they structure our perception of the world and our making sense of sensory data. He was not concerned with their origin, only their explanation for human experience. How these innate categories formed was never explained nor did he attempt to justify their innate quality.

Kant's fourfold grid of the analytic/synthetic divide created the puzzle of ampliative necessity. Logical empiricism attempted to protect it with verification; Quine's critique dissolved that strategy. Logical empiricism depended on a sharp analytic/synthetic boundary; Quine's critique eroded it. The classical reformulation is to distinguish epistemic routes (often *a posteriori*) from ontological grounds (*de re* necessity), thereby vindicating ampliative necessities. The classical philosophical solution distinguishes route of knowing between a naïve *a posteriori direct perception and the synthetic a priori* abstraction of necessity, universality, and causality grounded in being (*de re* necessity).

Kant's grid helped in that we see ampliative necessity is not derivable from enumeration alone. Yet it misleads when necessity is confined to phenomena through the mind's categories. Kant was not informed by empirical science or the molecular structure of molecules. Kant was simply at a disadvantage in forming his epistemological turn: his *Copernican Revolution*. On the other hand, the classical approach grants the *a priori* mark (necessity, universality) but relocates its ground in real natures—making synthetic *a priori* truths realist rather than idealist informed by empirical science and the structure of existence. A structure that is coherent and intelligible.

Kant's own general claim itself indicates a classical reformulation:

"Experience teaches that something is so; it never teaches that it cannot be otherwise."[26]

Precisely: experience alone does not yield necessity; but in the classical framework, experience occasions the intellect's grasp of **natures whose structures carry necessity**.

The classical reformulation moves from analytic rules from which necessity arises to ontological grounds. The analytic/synthetic map classifies sentences by how predicates relate to subjects conceptually. The classical reformulation classifies truths by their relation to being: *per se* truths that articulate what belongs to a nature (e.g., that a contingent being does not cause itself) and *per accidens* truths that track incidental conjunctions. The former supply ampliative necessities once the relevant nature is understood.

Why is this not merely induction in disguise? Induction aggregates cases, but the insight here is explanatory: from the analysis of motion as an act upon the potential, we see that an actualizer is required here and now to cause some x to become y. Similarly, from the intelligibility of being we see that contradictory attribution under the same aspect is impossible. These are not probabilities promoted to certainties, but necessities disclosed by the very nature of things. We have necessary causality and the necessity of the law of non-contradiction from nature itself insofar as the necessity and universality is abstracted from the way things necessarily and universally are. Experience occasions the intellect's grasp of **natures whose structures carry necessity and universality**.

Frege's emphasis on sense illuminates how identity statements can be informative; but necessity cannot be reduced to synonymy. Quine's holism warns us that no finite set of semantic rules will isolate the necessary. The classical approach accepts both lessons and adds a metaphysical anchor: essences ground necessity.

---

26 Die Erfahrung lehrt zwar, daß etwas so beschaffen sei; daß es aber nicht anders sein könne, lehrt sie nicht. — Kant, KrV A92/B124 (AA III).

But what is the payoff for science and metaphysics? In scientific practice, laws are not mere summaries of frequencies; they express dispositional structures. The classical picture explains why successful interventions reveal powers whose modal profile outruns observation. In metaphysics, it vindicates essentialist inference without collapsing into conceptual analysis. You might ask, isn't this simply rebranding analytic truths as 'essence', or as the 'nature of things', or the whatness of some x, what something is by its very ontological nature? the truths are not about words but about things; their necessity is explained by natures (e.g., the essence of water) rather than by inferential role. Doesn't Kripke already give us essences? The answer is yes—and the classical reformulation supplies the ontological backstory: *essences are forms grounding causal powers. What something is grounds the causal principles and first principles themselves given the specific context and time in which they exist as they are.*

To illustrate the reformulation within a classical synthetic *a priori* framework, we will show how the existential and universal quantifiers arise from experience by abstraction, and how they apply to *de re* modality in cases like 'Water is H2O'—aligning with Kripke's necessary *a posteriori* while addressing reference without possible worlds machinery.

First, we move from experience to quantification. Existential: repeated encounters ('this water, that water') ground the formation of a sortal concept 'water'. From this, the intellect affirms $\exists x\, \text{Water}(x)$. Universal: abstraction from many instances yields a universal nature; the intellect judges $\forall x(\text{Water}(x) \rightarrow F(x))$ for features F that belong *per se* to that nature.

"From experience the universal arises in the soul... from the many, one judgment."[27]

Thus quantifiers are not innate symbols but intellectual acts tracking being: $\exists$ (there is some) reflects recognition of instances; $\forall$ (for every) reflects the grasp of a universal nature.

---

27 ἐκ δ' ἐμπειρίας τὸ καθόλου ἐν τῇ ψυχῇ γίγνεται... — Aristotle, Posterior Analytics II.19, 100a5–6, ed. W. D. Ross (Oxford, 1949).

Formalizing the Reformulation:

Basic schema:

1. ∃x Water(x) [*a posteriori*: observation/ostension]
2. By abstraction, grasp the nature Water(–).
3. Discover *per se* feature E (essence): ∀x(Water(x) → E(x)).
4. Recognize *de re* necessity: □∀x(Water(x) → E(x)).

In the case of water, let E(x) := $H2O(x)$. Then once the essence is discovered empirically, the truth '∀x(Water(x) → $H2O(x)$)' expresses a *de re* necessity: it holds in virtue of what water is. Epistemically *a posteriori*; metaphysically necessary—just as Kripke maintains, but the possible world machinery is not needed. ***But why are possible worlds not needed?***

Possibleworlds semantics is a powerful modal logic occasionally used in our text, but classical metaphysics does not require it to explain modality. This is the second point, the truthmaker for the necessity operator '□φ' is the essence (form or nature) that grounds φ as belonging *per se* to its subject. Worlds can be used to regiment reasoning, but essence or nature of what something is suffices to explain why the modal operator is appropriate.

Let's compare Kripkean Reference against the new Synthetic *a-priori* Reformulated view of nature or essence. Kripke worries that descriptivism cannot fix reference. The classical approach agrees: reference is secured by an act of naming tied to the thing through causal contact and sustained usage—a causalhistorical chain. What makes the name latch onto an essence is not a description but the thing's form or nature or structure itself, to which the intellect has been introduced by experience. Let's use Kripke's example to illustrate this again but explain what rigid designators are for Kripke.

A rigid designator is a term that refers to the same object in every possible world in which that object exists. For example, 'Socrates' rigidly designates that very man, Socrates, in all possible worlds in which he exists. Even if Socrates had not been a philosopher or had died by hemlock, 'Socrates' would still pick out him, not someone else who fits a description.

By contrast, a non-rigid (or flaccid) designator can refer to different things in different possible worlds. For example, "The teacher of Aristotle," in the actual world, that description picks out Plato. But in another possible world, maybe Plato or someone else taught Aristotle – so the description would designate someone else.

For Kripke rigid designators in proper names such as "Socrates," "H20," and natural kinds (e.g. "water," "gold," "tiger") also function rigidly. Proper names refer directly, not via a description, to the same entity across possible worlds. In the case of natural kinds, "water" rigidly designates H20 in all possible worlds – even if in some other world people thought "water" was H30 or hydronium. This underlies Kripke's claim that statements like "Water is H2O" are necessary *a posteriori* truths. *A posteriori* because we discover it empirically and necessary because if "water" refers rigidly to H20, then in every world where water exists, it must be H2O.

Rigid designators break with earlier theories such as Frege, where names were often treated as shorthand for definite descriptions. A definite description is a denoting phrase in the form of x where the x is a noun phrase or a singular common noun. For Kripke, if names were equivalent to descriptions, they would not be rigid – because the description could pick out different individuals in different possible worlds. Kripke argues instead that names and some natural kind terms "stick" to their referents independently of contingent descriptions. In other words, certain names and phrases refer to particular things independent of their descriptions. Like "water" rigidly designates H2O in all possible worlds. Therefore the phrase by virtue of "water" being a rigid designator, "Water is H2O" are necessary *a posteriori* discovered empirically but necessary because "water" refers rigidly to H20, such that in every world where water is exists, it must be H2O. Note water is essentially H2O, but it is necessarily H2O not because of what water is essentially but because water rigidly designates water as H2O across all possible worlds. And that is the key difference between Kripke and classical metaphysics. Both make essentialist claims, but Kripke does so by means of rigid designators and all possible worlds. The classical approach simplifies this and says the nature of things themselves are essentially and necessarily what they are. And from this, the intellect abstracts the universality and necessity from the whatness, form, or nature, or structure of things. For Kripke, a rigid designator is a term that designates the same object or kind in every possible world in which that object or kind exists, unlike descriptive terms which may vary across worlds. Let's illustrate the difference between rigid designators by stipulation vs. by what something is.

Illustration with 'Water is H2O'

Let 'Water' be a rigid designator of the natural kind whose essence includes being H2O.

Then:

A. $\exists x\, \text{Water}(x) \wedge \text{Observed}(x)$.
B. Empirical discovery: $\forall x(\text{Water}(x) \rightarrow H2O(x))$.
C. Classical necessity: $\Box \forall x(\text{Water}(x) \rightarrow H2O(x))$ (true in virtue of essence, not by stipulation).

The role of $\exists$ and $\forall$ is thus grounded in experience driven abstraction; the role of $\Box$ is grounded in essence. No possible worlds ontology is needed to underwrite the claim. Therefore, the possible world semantics is unnecessary.

9.

FREGE AND KRIPKE: FORMAL AND MODAL RECONSTRUCTIONS OF AQUINAS' THIRD WAY

"The sense of a proper name is the mode of presentation of the object."[28]

Frege clarifies informative identity; Kripke shows some identity statements are necessary *a posteriori due to rigid designators*. Aquinas's Third Way can be cast in first-order and modal idioms to display the necessity of a non-derivative cause without presuming analyticity. To show the reformulated synthetic *a priori* in action, consider Aquinas' Third Way (from possibility and necessity) in a Fregean and a Kripkean idiom.

Fregean first-order skeleton (without modality). Let Ex = 'x exists (here and now)', Cont(x) = 'x is contingent', Dep(x) = 'x's existence depends *per se* on another', and Caus(y,x) = 'y is an essentially *per se* cause of x':

1. $\exists x\, Cont(x)$
2. $\forall x(Cont(x) \rightarrow Dep(x))$ [synthetic *a priori*: contingency entails dependence]
3. $\forall x(Dep(x) \rightarrow \exists y(Caus(y,x) \wedge y \neq x))$
4. No infinite *per se* regress of causes (well-foundedness).

$\therefore \exists z\, Necessary(z)$ and $(\neg Dep(z))$ — a non-dependent existent ('necessary' in Aquinas' sense).

---

28 Der Sinn eines Eigennamens ist die Art des Gegebenseins des Gegenstandes. — Gottlob Frege, "Über Sinn und Bedeutung," Zeitschrift für Philosophie und philosophische Kritik 100 (1892): 28–29.

Quantifiers here arise from experience (1), are extended by abstraction to universal claims about dependence (2), and yield a metaphysically necessary conclusion about a nondependent being—without appeal to possible worlds, because Necessary '□' is read as 'true in virtue of essence'.

Modal enrichment (S4/S5-style) using modal operators.

Let $N(x) := \Box Ex$ and $Cont(x) := \Diamond \neg Ex$.

A1. $\exists x\, Cont(x)$

A2. $\forall x(Cont(x) \rightarrow \Diamond \exists y\, Caus(y,x))$ [potency needs act — synthetic *a priori*]

A3. $\Box \neg Inf$ *Per se*   [no infinite essential regress is possible]

A4. $\forall x \forall y(Caus(y,x) \rightarrow (Ex \rightarrow Ey))$ [existential priority of causes]

∴ $\exists z\, \Box\, Ez$

Explanation: empirical encounter with contingent beings (A1) occasions the intellect's grasp of a necessary structural truth (A2). Together with (A3) and (A4), we infer a being whose existence does not depend *per se* on another. The conclusion's necessity is metaphysical (*de re*), while our epistemic route may be *a posteriori*—this matches Kripke's necessary *a posteriori* pattern.

Fregean analyticity is not at stake: the argument's key steps are ampliative about being and causation. Yet they are not mere inductions; rather, they are synthetic *a priori* as a reformulation in the classical sense—truths seen as necessary once being, potency, and act are properly understood.

Addressing Kripke's concerns more directly about reference, rigid designation works because names are anchored in things via causal origins; on a classical reading, those anchors succeed because forms or nature of things are the principles by which things are knowable and nameable. As far as necessity, the necessary *a posteriori* is intelligible because the route of discovery can be empirical while the ground of truth is essential. As far as possible worlds, their utility remains, but they do not ground modality; essences do.

1. $\exists x\ Cont(x)$;
2. $\forall x(Cont(x) \rightarrow Dep(x))$;
3. $\forall x(Dep(x) \rightarrow \exists y\ Caus(y,x))$;
4. No infinite *per se* regress; $\therefore \exists z\ \neg Dep(z)$.

Modal enrichment (S4/S5):

A1. $\exists x \Diamond \neg Ex$;
A2. $\forall x(\Diamond \neg Ex \rightarrow \Diamond \exists y\ Caus(y,x))$;
A3. $\Box \neg Inf$ *Per se;*
A4. $\forall x \forall y(Caus(y,x) \rightarrow (Ex \rightarrow Ey))$; $\therefore \exists z\ \Box\ Ez$.

Quantifiers and modality can be naturalized within classical metaphysics: $\exists$ and $\forall$ track the intellect's ascent from particulars to universals; $\Box$ tracks essence. This yields Kripkeanfriendly necessities without a reliance on possibleworlds ontology, and it integrates Aquinas's Third Way into contemporary formal perspectives. This reformulation allows all the five ways, and empirical science to be integrated into contemporary formal logic and modality.

10.

RESOLUTION: CLASSICAL METAPHYSICS BEYOND THE ANALYTIC–SYNTHETIC DIVIDE

"Being and truth are convertible."[29]

Returning necessity to being turns the analytic–synthetic divide into a taxonomy of epistemic routes rather than a boundary on necessity. Identity, non-contradiction, and causality function as first principles seen by intellect in light of being; each is universal and necessary because of what things are.

The analytic–synthetic problem dissolves once necessity is returned to being. Identity, non-contradiction, and causality are first principles discovered by abstraction from particulars and recognized as universally and necessarily true because of the nature of being.

Identity, non-contradiction, and causality stand as first principles discovered by abstraction from experience and recognized as necessary because they express structures of being. This rehabilitates causal explanation against Humean skepticism, reformulates Kant's synthetic *a priori* without idealism, and integrates Quine's and Kripke's insights within a realist metaphysics.

The classical account of abstraction delivers universal and necessary principles from experience without collapsing into subjectivism or conventionalism. It resolves the issues of rigid designators and possible world semantics. Likewise, it resolves the analytic/synthetic debate

---

29 Ens et verum convertuntur. — Thomas Aquinas, Summa Theologiae I, q.16, a.3 (Leonine ed., vol. 4).

and the tension between rigid designators and descriptive approaches. Reformulating the synthetic *a priori* using classical metaphysics and epistemology clarifies Kant's insight while repairing its restriction to phenomena. Frege's and Kripke's concerns are accommodated: logic is not the whole of necessity, and some necessities are known *a posteriori*. The intellect discovers, rather than imposes, the structures of being consistent with observation and consistent with the applied scientific disciplines. In effect, the reformulation has answered David Hume's critique on the empirical sciences and causality itself. It has demonstrated that Kant's Transcendental Idealism was an unnecessary turn. And the analytic turn by Frege and the possible world semantics of Kripke although having clear utility are unnecessary for empirical science, metaphysics, epistemology, and modal logic when it comes to possibility and necessity abstracted from being itself.

What follows are various arguments using the Classical Reformulated Synthetic *a priori* approach to modal logic and formal semantics. These are not mere examples, these are in fact metaphysical explanations using formal modal logic grounded in the necessity and universality of being itself.

11.

THE ESSENCE AND FORM OF WATER: A CLASSICAL METAPHYSICAL GROUND IN CHEMISTRY

How do we know what is? What are the epistemic conditions by which a knowing subject can know that some substance is water and some other substance is not? And how does metaphysics inform and determine the epistemic conditions by which a knowing subject can know what the essence of some substance is? This following explores the essence of water through the lens of Classical metaphysics and modern molecular.chemistry. While water is scientifically described as H_2O, this formula alone does not capture what makes water the substance that it is. We argue that the essence of water is not reducible to its chemical formula nor to its subatomic constituents, but is found in the molecular . form—the quantum-mechanical configuration that unifies hydrogen and oxygen atoms into a single substance with unique properties. By integrating the Classical concepts of form, matter, and essence with molecular chemistry, we show how water's identity emerges only at the level where its form actualizes matter into a cohesive whole.

The question of what makes water 'water' invites both scientific and philosophical reflection. From a scientific perspective, water is typically described as H_2O, a molecule composed of two hydrogen atoms and one oxygen atom. From a philosophical perspective, particularly within the Classical metaphysical tradition, we must ask: what is the essence of water? What is the form that unifies its parts into a single substance with a specific nature? This paper addresses these questions by integrating molecular chemistry with Classical metaphysics, distinguishing between the formula H_2O, the form that organizes it, and the essence that arises from form realized in matter.

H_2O consists of two hydrogen atoms and one oxygen atom covalently bonded at a specific angle (~104.5°). If you separate these atoms, you no longer have water—you have hydrogen gas (H_2) and oxygen gas (O_2) as distinct substances. The specific molecular structure is essential to water's properties, such as hydrogen bonding, polarity, and its anWhen hydrogen and oxygen atoms are considered separately, the identity of water is lost. At the atomic level, there are only protons, neutrons, and electrons. At the subatomic level, these particles are excitations of quantum fields, where no trace of what makes water distinct remains. Water's identity does not reside at these levels; it emerges only when these particles are arranged into the molecular form of H_2O. omalous behaviors.

From the perspective of molecular chemistry, the unifying principle that makes two hydrogen atoms and one oxygen atom a single H_2O molecule is the quantum-mechanical arrangement of electrons that stabilizes the covalent bonds and enforces a specific geometric configuration. This pattern determines the molecule's polarity, hydrogen-bonding capacity, and ultimately the specific nature of water as a substance. This arrangement is the activating principle or form that organizes matter into a cohesive whole as water.

In Classical metaphysical terms, form is the activating and unifying principle that actualizes matter into a substance of a specific kind. Two hydrogen atoms and one oxygen atom are only potentially water; they require the activating principle—the molecular form—to become actually water. The activating quantum-mechanical configuration is the form of water, and the essence of water is the form-in-matter—the H_2O structure realized in its proper material components. Thus, the essence of water is not simply the chemical formula H_2O, nor the abstract configuration, but the configuration as instantiated in hydrogen and oxygen to make a unified substance.

For Classical metaphysics, form actualizes matter, and essence is the form as realized in matter. Form (eidos) is the activating principle; matter (hylē) is the potential substrate; essence (to ti ēn einai) is the form-in-matter that makes something what it is. Therefore, the essence of water is not just the parts (hydrogen and oxygen), nor the bare formula, but the actualized structure—the molecular form instantiated in its proper matter. The essence of water cannot be reduced to its formula or to its atomic parts. It resides in the specific molecular structure H_2O, where the form—the quantum-mechanical arrangement—actualizes matter into a unified substance with unique properties. This integration of Classical metaphysics and molecular chemistry clarifies that water's identity emerges only when the proper matter is organized by its activating form. Thus, the essence of water is the form-in-matter, the H_2O molecule as an organized whole that makes water what it is and not something else.

12.

THE ONTOLOGICAL GROUNDING AND THE INTELLIGIBILITY OF H₂O: A MODAL ARGUMENT AGAINST APOPHENIA

The recognition of coherence, regularity, and intelligibility in natural phenomena raises the question of whether such order is intrinsic to the world or merely a projection of the human mind. Ever since the critique of Hume, Kant's epistemic turn, and W. V. O. Quine – "On What There Is" (1948) foundational essay challenged traditional ontological categories and the analytic/synthetic distinction, directly questioning the meaningfulness of much metaphysical discourse, metaphysics became logically rigorous in its persuit to demonstrate ontological or metaphysical necessity. David Lewis – On the Plurality of Worlds (1986) in his influential book developed a comprehensive metaphysical system based on modal realism, defending the existence of possible worlds as concrete realities. It became a centerpiece in current analytic metaphysics. The followwung goes further and argues that the intelligibility observed in H_2O is not an instance of apophenia but a reflection of its ontologically grounded form and essence. Using a quantified modal framework, we demonstrate that the regularity in H_2O is intersubjectively verifiable, predictively robust, and causally explicable, thereby necessitating its grounding in the essence of the substance rather than in mental constructs. This conclusion aligns with classical metaphysics, where intelligibility is a real property of being, accessible to human intellect.

Human cognition has a natural tendency to seek patterns. In some cases, this leads to apophenia—the erroneous perception of order in randomness. However, in the case of natural substances such as water (H_2O), the regularities observed are neither illusory nor subjective. The ordered behavior of H_2O—its molecular geometry, anomalous density properties, hydrogen bonding network, and phase transitions—forms the basis for successful scientific predictions and explanations.

This raises a fundamental metaphysical question: Are the patterns we perceive in H_2O merely projections of the human mind, or do they reflect an objective, ontologically grounded order inherent in the substance itself? Following classical metaphysics, we argue that the regularity and intelligibility of H_2O are grounded in its essence (form). This grounding is demonstrated through a rigorous modal logical argument, which shows that:

- The order in H_2O is empirically verifiable and intersubjectively confirmed.
- Apophenia, by definition, lacks such verification and causal grounding.
- Therefore, the order in H_2O cannot be reduced to mental projection but must be grounded in its essence.

Apophenia refers to false pattern recognition. It arises in contexts where patterns lack objective causal structure and are not reproducible across observers. Such patterns fail the tests of empirical and intersubjective verification. Apophenia reduces the ontological nature of things, or the ontological state, or state of affairs to the epistemological conditions of a Knowing subject. It limits existence to perception and phenomenological experience and Inference.

Classical metaphysics posits that natural substances have forms or essences that ground their behaviors and properties. This order is not constructed by the observer but discovered as a real feature of being. To express these distinctions rigorously, we employ quantified modal logic. This allows us to differentiate between what is necessarily true (□), possibly true (◇), and contingently true, capturing the metaphysical relationships between order, essence, and intelligibility.

Symbolization Key

□P: P is necessarily true
◇P: P is possibly true
I(x): x is intelligible
R(x): x exhibits real regularity
G(x): x's order is grounded in its essence or form
Pj(x): x's order is merely a mental projection
E(x): x exists
F(x): x has an essence or form
H: H_2O
M: Intellect or human mind

Premises
1. □∀x(I(x) → (R(x) → ◇M understands x))

Necessarily, if x is intelligible and exhibits real regularity, then it is possible for the intellect to understand x.

2. □∀x(R(x) → (G(x) ∨ Pj(x)))

Necessarily, if x exhibits regularity, that regularity is either grounded in its essence or is a mere mental projection.

3. □∀x(Pj(x) → ¬E(x) ∨ ¬(R(x) is intersubjectively verifiable))

Necessarily, if regularity is merely a projection, either x does not exist independently, or its regularity is not intersubjectively verifiable.

4. □∀x(G(x) → (F(x) ∧ E(x)))

Necessarily, if x's regularity is grounded in its essence, then x has an essence and exists.

5. R(H) ∧ E(H)

H_2O exhibits regularity (e.g., molecular structure) and exists.

6. R(H) is intersubjectively verifiable

The regularity of H_2O is empirically observed and confirmed across observers.

Derivation

7. From (5) and (6), using (3): $\neg Pj(H)$:

H_2O's regularity is not a projection.

8. From (5) and (2): $R(H) \rightarrow (G(H) \lor Pj(H))$. Since $\neg Pj(H)$, we have $G(H)$.

9. From (8) and (4): $G(H) \rightarrow (F(H) \land E(H))$. Therefore, $F(H) \land E(H)$.

H_2O has an essence and exists.

10. From (1), with I(H) following from $(R(H) \land G(H))$, we have $\Diamond M$ understands H.

It is possible for the intellect to understand H_2O because it exists and has an essential nature, the composition and unity of two hydrogen atoms and one oxygen atom.

Conclusion from the Logic:

$\Box(R(H) \land E(H) \land F(H) \land G(H) \land \Diamond M \text{ understands H})$

Therefore, it is necessarily true that H_2O's regularity is real, grounded in its essence, and intelligible to the intellect.

This modal argument demonstrates that the observed order in H_2O is not a case of apophenia but reflects a real, ontologically grounded essence. Apophenia lacks empirical and intersubjective verification, whereas H_2O's structure is confirmed through robust scientific investigation. Moreover, the ability of the intellect to grasp and predict the behavior of H_2O presupposes that the order is not merely subjective. This finding is consistent with a metaphysical framework where intelligibility is a real property of being, not an imposed fiction.

13.

FROM THE INTELLIGIBILITY OF H₂O TO THE NECESSITY OF A NECESSARY BEING

The ordered structure and intelligibility of H_2O (water) exemplify the metaphysical principle that contingent regularities require grounding beyond themselves. This chapter argues that the real, intersubjectively verifiable order in H_2O cannot be explained by an infinite regress of contingent causes. Using quantified modal logic and the Principle of Sufficient Reason, it is demonstrated that the intelligibility of H_2O points toward the existence of a necessary being that grounds the order of all contingent beings include H_2O.

Water (H_2O) exhibits remarkable regularities: its molecular geometry, hydrogen bonding, and anomalous physical properties make life possible and are robustly confirmed by scientific investigation. These regularities raise the question of whether the order we observe is self-explanatory or requires grounding in something more fundamental. This paper applies the Principle of Sufficient Reason (PSR) and modal logic to argue that the intelligibility of H_2O implies the existence of a necessary being as the ultimate ground of order.

A contingent being is one that exists but could have failed to exist under different conditions. A necessary being, by contrast, exists in all possible worlds and cannot fail to exist. The intelligibility of contingent beings ultimately requires grounding in a necessary being.

The Principle of Sufficient Reason (PSR) states that every contingent fact has an explanation. Applied to the regularity of H_2O, this means that its existence and properties require a sufficient reason beyond themselves. An infinite regress of contingent explanations fails to satisfy this principle, leading to the conclusion that there must be a necessary being that grounds all contingent intelligibility.

Argument from the Intelligibility of H_2O

Premises

1. R(H): H_2O exhibits real regularity and intelligibility.
2. $\Box \forall x(R(x) \rightarrow$ (x is contingent $\lor$ x is necessary)).
3. C(H): H_2O is contingent because it depends on physical laws and conditions that could have been otherwise.
4. $\Box \forall x(C(x) \rightarrow \exists y(G(y,x)))$: Necessarily, every contingent being requires an explanation or ground.
5. $\Box \neg \exists z(\forall x(C(x) \rightarrow G(z,x)) \land C(z))$: No contingent being explains all contingent beings.
6. $\therefore \exists N(N$ is necessary $\land \forall x(C(x) \rightarrow G(N,x)))$: There exists a necessary being N that grounds all contingent beings, including H_2O.

Another Formulation:

Let H denote H_2O, C(x) denote 'x is contingent', N(x) denote 'x is a necessary being', and G(y,x) denote 'y grounds x'.

1. $R(H) \land C(H)$
2. $\Box \forall x(C(x) \rightarrow \exists y(G(y,x)))$
3. $\Box \neg \exists z(\forall x(C(x) \rightarrow G(z,x)) \land C(z))$
4. $\therefore \exists n(N(n) \land \forall x(C(x) \rightarrow G(n,x)))$

H_2O serves as an illustrative case of how contingent intelligibility points beyond itself to a necessary cause. The regularity and order of water are contingent; they depend on specific physical laws and conditions. By the PSR, contingent intelligibility requires an ultimate explanation, and an infinite regress of contingent causes fails to provide one. Thus, the intelligibility of H_2O leads to the conclusion that there exists a necessary being that grounds the order of all contingent realities.

14.

ONTIC STRUCTURAL REALISM, QUANTUM GRAVITY, AND THE ARGUMENT FOR A FIRST EFFICIENT CAUSE

Developments in quantum gravity and ontic structural realism (OSR) challenge traditional substance-based metaphysics by privileging relations and informational structures over material substrates. This paper argues that such structural accounts do not undermine the classical framework of metaphysical causality but rather demand a deeper grounding in being. Specifically, we contend that the apparent indeterminism and ontological relationality of quantum phenomena—particularly in the context of quantum superpositions—point to a hierarchy of contingent actualizations that cannot be self-explanatory. By integrating insights from OSR and quantum gravity into classical metaphysics, we defend the necessity of a First Efficient Cause: a purely actual, non-contingent being that grounds the intelligibility and existence of all relational structures. The argument is formalized using symbolic logic and modal ontology to show that even in a fundamentally relational quantum universe, the principle of causality entails the existence of a necessary being.

The interpretation of quantum mechanics has long raised metaphysical questions concerning the nature of causality, potentiality, and actuality. Recent work in quantum gravity and ontic structural realism intensifies these issues by reconceptualizing the basic constituents of reality not as particles or fields, but as networks of relations or informational structures. Yet, this move away from substance ontology does not evade the question of ontological grounding—it deepens it. Far from being rendered obsolete, classical metaphysics provides the scaffolding necessary to understand what it means for such structures to exist. In particular, we defend the argument for a First Efficient Cause within the context of quantum ontology.

Quantum superpositions suggest that systems can exist in multiple potential states until an act of measurement results in a definite outcome. While interpretations differ, all agree that quantum systems evolve according to well-defined mathematical rules (e.g., the Schrödinger equation), implying a structured, albeit non-classical, form of causality.

Research in loop quantum gravity (LQG), causal set theory, and string theory proposes that spacetime itself is emergent from more fundamental, discrete, or informational structures.

Ontic Structural Realism (OSR) holds that the fundamental ontology of reality consists not in objects but in the relational structure among them. This aligns well with quantum mechanics, where particles lack definite properties independent of measurement contexts.

Yet, these structures still participate in being—and being, per the classical view of metaphysics, requires grounding in actuality. The relational structure is not self-explanatory.

In classical metaphysics:

- Potency: That which can be, but is not yet.
- Act: That which is.
- Efficient Cause: That which brings something from potency to act.

A contingent structure—be it material or relational—cannot actualize itself. This leads us to the necessity of a First Cause, which is Pure Act (*actus purus*), i.e., a being whose essence is its existence (*ipsum esse subsistens*).

The Formal Argument

Let:

- C(x) = 'x is a contingent being'
- A(x) = 'x is actualized'
- P(x) = 'x is in potency'
- E(x, y) = 'x is efficiently caused by y'
- N(x) = 'x is a necessary being'
- G = the necessary First Efficient Cause

Axioms:

1. $\forall x \, [C(x) \rightarrow \exists y \, E(x, y)]$
2. $\forall x \, [P(x) \rightarrow \exists y \, (A(y) \wedge E(x, y))]$
3. $\neg \exists x \, [E(x, x)]$
4. $\neg \exists x \, [C(x) \wedge \forall y \, (E(x, y) \rightarrow C(y))]$
5. $\exists x \, [N(x) \wedge \forall y \, (E(y, x) \rightarrow y = x)]$

Conclusion:

$\therefore \exists G \, [N(G) \wedge \forall x \, (C(x) \rightarrow \exists y \, (E(x, y) \wedge (y = G \vee E(y, G))))]$

Even in a fundamentally relational or probabilistic universe:

- Superpositions represent potentialities, not pure being.
- Quantum events (measurements, decoherence) represent actualizations.
- These actualizations demand an ontological ground—a being that is not itself a composite of potentials.

Quantum physics, far from displacing classical metaphysics, invites its revival on deeper terms. Ontic structural realism and quantum gravity reveal a universe rich in structure but poor in ontological independence. Only by grounding this structured potentiality in a purely actual, necessary being can we account for its intelligibility and existence. The First Efficient Cause remains not only philosophically necessary, but scientifically relevant.

15.

THE METAPHYSICAL NECESSITY OF A SELF-SUBSISTENT, INCORPOREAL, AND INFINITE BEING

Previous chapters argued for a defense of classical metaphysics and the necessity of a Necessary First Cause of all existence. This chapter goes further and argues that classical metaphysics necessitates the existence of a Being whose essence and existence are identical and who is therefore self-subsistent *per se*. This Being must also be incorporeal and infinite to serve as the necessary ground and ultimate cause of all contingent beings. Drawing from key texts, and employing both symbolic and modal logic, the chapter demonstrates the metaphysical necessity of such a Being. The analysis addresses the limitations of modern physics in accounting for metaphysical foundations and contrasts these with the rigorous ontological framework provided by classical metaphysics.

Classical metaphysics begins with the question of being *qua* being (ἐπι τὸ ἀν ἐνόντι η ἐπιστήμη). The analysis of substance, causality, and the act-potency distinction leads to the conclusion that contingent beings require a cause whose existence is not itself contingent. This first cause must be, by necessity, a Being whose essence is existence itself *(ipsum esse subsistens)*.

Let $C(x)$ denote x is contingent; $E(x)$ denote x exists; and $S(x)$ denote x is self-subsistent (i.e., its essence = existence).

(1) $\exists x\, C(x) \rightarrow \exists y\, [\neg C(y) \wedge Cause(y,x)]$

(2) $\neg C(y) \Rightarrow S(y)$

(3) $S(y) \Rightarrow E(y) \wedge Essence(y) = Existence(y)$

Given the reality of contingent beings $(C(x))$, there must exist a non-contingent cause, that is self-subsistent, whose essence is its existence. Classical metaphysics maintains that corporeal beings are composites of matter and form (hylomorphism), thus having potency. A purely actual being (*actus purus*) cannot have potency, and thus cannot be material. This can be frames as follows:

Let $M(x)$ denote x is material; $P(x)$ denote x has potency.

(4) $M(x) \Rightarrow P(x)$

(5) $\neg P(y) \Rightarrow \neg M(y)$

(6) $S(y) \Rightarrow \neg P(y)$

Therefore, a self-subsistent being cannot be material: it must be incorporeal. Classical metaphysics distinguishes between the infinite in act and the infinite in potential. A finite being is limited by form and matter. A being whose essence is existence is unlimited, lacking all potency and thus infinite. Formulated thus:

Let $F(x)$ denote x is finite; $I(x)$ denote x is infinite.

(7) $S(x) \Rightarrow \neg F(x) \Rightarrow I(x)$

A self-subsistent being must therefore be infinite.

Adding modality to the argument:

Let $\square$ denote necessity and $\lozenge$ denote possibility. For contingent beings:

(8) $C(x) \Rightarrow \lozenge E(x) \wedge \neg \square E(x)$

For the necessary being:

(9) $S(x) \Rightarrow \square E(x)$

Thus, the self-subsisting necessary being exists in all possible worlds, whereas contingent beings do not.

Modern physics describes the behavior of matter and energy within spacetime but does not account for why there is something rather than nothing. The limitation of modern physics is its lack of explanatory power. It is not that science has not yet discovered the answers and pushed the boundaries. It simply cannot explain, and it is not equipped to explain why there is something rather than nothing and why that something is *intelligible* and *coherent*. Why is there uniformity, causality, and regularity throughout nature rather than non-uniformity, irregularity, random events without causality. The Quantum paradox of the uncertainty principle (i.e., what Einstein called the ghost in the box) only heightens the question, why is there uniformity and regularity at the molecular level when there is possible uncertainty at the quantum level.

Asking science to explain such questions is like demanding a microwave to produce ice. Asking science to explain such questions is asking science to engage in doing metaphysics, which it's not equipped to do; and that is exactly what theoretical physics has become. But the classical framework, through the analysis of causality and being, transcends empirical constraints and addresses the metaphysical foundation that physics presupposes but cannot explain.

Being itself demands a first cause and leads necessarily to the affirmation of a Being that is self-subsistent, incorporeal, and infinite. This Being is not merely the terminus of an explanatory chain but the very ground of being itself. A bold claim shall be argued. By implication: a) science is not equipped to answer the most fundamental question of what caused the universe and b) existence, as we know it, is contingent and could not have always existed and could not have come into existence of its own accord. The next chapter will argue the reason why the universe is intelligible and coherent is because of an intelligent and necessary subsistent First Cause that is both intelligible and coherent.

16.

THE NECESSITY OF DIVINE BEING ACROSS ALL POSSIBLE WORLDS

This chapter argues that the very possibility of any being presupposes an intelligible essence, and that intelligibility itself requires grounding in a necessary intellect. By analyzing the conditions for the possibility of beings—namely, their formal and final causes—we demonstrate that contingent existents cannot account for their own intelligibility. Only a self-subsistent, uncaused cause whose essence is identical with its act of existence can ground the real distinction between essence and existence in all other beings. Thus, a necessary First Cause—pure act and divine intellect—is required as the ontological ground of both possibility and actuality across all possible worlds.

1. In classical metaphysics, a thing is possible only if it is intelligible (i.e., has a non-contradictory essence or form or nature).

2. Intelligibility implies a formal cause, which implies that the thing is grounded in a mind (the Divine Intellect).

3. Contradiction—e.g., a square circle or a unicorn with a triangle-shaped round horn—is unintelligible, not because it is merely unusual, but because it violates the principle of non-contradiction and therefore has no form or essence capable of grounding in any mind.

4. Thus, only that which is non-contradictory can be conceived by the Divine Intellect and thereby possess the condition for possible existence.

5. Nature, therefore, must be ordered and intelligible. Its laws must reflect regularity, proportion, and finality because anything otherwise (e.g., pure chaos or contradictory states of affairs) could not be grounded in divine reason.

6. Contingent beings have a real distinction between essence and existence and thus require a cause for their act of being, rooted in the necessary act of divine understanding. Therefore, in all possible worlds, God exists necessarily.

To ensure metaphysical rigor, we clarify and defend the Axioms underlying the modal argument:

A1. Intelligibility is a precondition for real possibility: $\Diamond E(x) \rightarrow I(x)$

A2. Intelligibility implies an internally coherent form or essence (non-contradiction): $I(x) \rightarrow \neg(Ess(x) \wedge \neg Ess(x))$

A3. Forms and essences exist as divine ideas: $Ess(x) \rightarrow D(x)$

A4. Divine ideas exist only in a necessarily existing, self-subsistent, intelligent mind: $D(x) \rightarrow \exists y\ (M(y) \wedge S(y) \wedge N(y) \wedge G(x, y))$

A5. The only being for whom essence = existence and who is necessary, self-subsistent, and intelligent is God: $\forall y\ (M(y) \wedge S(y) \wedge N(y) \wedge E(y) = Ess(y)) \rightarrow y = God$

A6. Therefore, all possibility and intelligibility across possible worlds presupposes the necessary existence of God: $\forall W\ \exists y\ (M(y) \wedge S(y) \wedge N(y))$

W = A possible world
B = A being in a world
C(x) = x is contingent
N(x) = x is necessary
I(x) = x is intelligible
D(x) = x exists as a divine idea
E(x) = x exists
Ess(x) = essence of x
G(x, y) = x is grounded in y

S(x) = x is self-subsistent

M(x) = x is an intelligent mind

Formal Argument with the above Integrated Axioms:

1. $\forall W\ \forall B{\in}W\ (\Diamond E(B) \rightarrow I(B))$ [From A1: Possibility requires intelligibility]

2. $\forall B\ (I(B) \rightarrow \neg(Ess(B) \wedge \neg Ess(B)))$ [From A2: Intelligibility implies non-contradiction]

3. $\forall B\ (I(B) \rightarrow Ess(B))$ [From A2 and logic: If intelligible, then has essence]

4. $\forall B\ (Ess(B) \rightarrow D(B))$ [From A3: Every essence exists as a divine idea]

5. $\forall B\ (D(B) \rightarrow \exists x\ (M(x) \wedge S(x) \wedge N(x) \wedge G(B, x)))$ [From A4: Divine ideas require a necessary intelligent mind]

6. $\forall x\ (M(x) \wedge S(x) \wedge N(x) \wedge E(x) = Ess(x)) \rightarrow x = God$ [From A5: Only God fits this description]

   $\therefore \forall W\ \exists x\ (M(x) \wedge S(x) \wedge N(x))$ [From A6 and (1)–(6): In all possible worlds, God exists necessarily]

God, as *ipsum esse subsistens* and the divine intellect, is the necessary and sufficient condition for the possibility and intelligibility of all beings across all possible worlds. The impossibility of contradictions further confirms that the ground of being must be pure act, without potency or internal opposition. Contradictory constructs—like square circles or unordered realities—are not merely impossible in practice but metaphysically impossible because they have no grounding in a rational divine intellect. Regularity and order in nature are not contingent features but flow necessarily from the principle that intelligibility is a prerequisite of existence. The laws of nature are reflections of divine reason, not arbitrary patterns. What science cannot explain, classical metaphysics provides an inference to a best explanation (IBE) grounded in being itself and demonstrated by modal logic and empirical science. The following summary is explained in detail in the appendix on the metaphysics of Divine Ideas.

- Classical metaphysics grounds all modal possibility in divine intelligibility.

- Only that which has an internally coherent form can exist even potentially, and that coherence requires a divine intellect to conceive it.

- Contradictory "objects" are not grounded in being and thus are not possible.

- Natural order reflects divine intelligibility; irregular or chaotic alternatives are not metaphysically possible. A tornado appears chaotic, but it follows natural laws even though it is highly unpredictable. It is these laws that introduce regularity in chaos.

- An intelligent being must exist in all possible worlds to ground the intelligibility of any possible entity.

- Therefore, in every possible world, God must exist as the necessary, self-subsistent, intelligent cause of all that is possible and actual.

- This is not a mere modal hypothesis but a metaphysical deduction from the nature of being and the structure of intelligibility itself.

17.

FINAL CAUSALITY AND THE INTELLIGIBILITY OF NATURE: A DEFENSE OF FINAL CAUSALITY

This chapter offers a philosophical defense of final causality within the classical philosophical tradition, arguing that purpose or teleology is essential to the intelligibility of nature. Through an analysis of natural regularity, causal explanation, and metaphysical structure, we show that final causes are not dispensable pre-scientific relics but necessary components of a coherent metaphysics.

Just as a chair is composed of various parts to form what it is to be a chair, its final end or telos determines why it is composed the way it is rather than in some other way. Formal causality explains what it is while final causality explains the why. A chair exists for one to sit, and the structure explains its nature.

The case study of why hydrogen and oxygen combine to form water rather than fire serves to illustrate the interrelationship between final and formal causes and to reveal the metaphysical depth often obscured by reductionist frameworks. A symbolic formalization of the argument is included via the S5 modal logic schema. The argument is also revised to consider the possibility that not all natural events exhibit intrinsic directedness. A further illustration is provided using genetic adaptation apart from evolutionary mechanisms.

Aristotle's four causes—material, formal, efficient, and final—remain foundational to classical metaphysics. While modern science has largely abandoned final causality as methodologically unnecessary, the classical tradition maintains that final causes are indispensable to a coherent understanding of the natural world. This paper defends final causality as a real and irreducible aspect of nature, necessary for explaining the directedness of most natural processes.

17.1 The Four Causes Revisited

- Material cause: what something is made of
- Formal cause: what something is, its essence
- Efficient cause: what brings it about
- Final cause: the end or purpose for which it exists

According to classical metaphysics, the final cause is "the cause of causes," because efficient causality presupposes directedness toward an end.

Revised Argument for Final Causality (Verbal Form)

Premise 1: Many natural substances exhibit consistent, goal-directed behavior (e.g., acorns become oak trees).

Premise 2: This behavior implies regularity not imposed externally but intrinsic to the substance.

Premise 3: Intrinsic regularity implies a directedness toward specific ends.

Premise 4: Directedness toward specific ends constitutes final causality.

Conclusion: Therefore, final causality is a real and irreducible component of natural explanation, at least in most cases.

It may be argued that not all natural phenomena exhibit directedness (e.g., radioactive decay or quantum tunneling). However, such exceptions do not nullify the general intelligibility that final causality offers for a wide range of natural phenomena. The classical view need not claim universal application without qualification, but rather that final causality is essential wherever intelligibility, structure, and teleological behavior are observed.

17.2 Formalization in Modal Logic (S5)

Let:

Cx: x is a contingent natural entity
Dx: x exhibits directed behavior
Fx: x has a final cause
NEC: necessity ($\square$), POS: possibility ($\lozenge$)

1. $\exists x(Cx \wedge Dx)$ (Some contingent natural things exhibit directedness)
2. $\forall x(Dx \rightarrow Fx)$ (Directedness entails a final cause)
3. $\exists x(Cx \wedge Fx)$ (Therefore, some contingent things have a final cause)
4. $\exists x(Cx \wedge \exists y(\square Fy))$ (Some contingent x implies some necessarily final cause y)

It is acknowledged that not all natural events are directional, but the core metaphysical insight remains. Final causality is a real and irreducible component of natural explanation, at least in most cases.

17.3 Illustration: Why Water Is Not Fire

Consider the combination of hydrogen and oxygen to form water. Scientifically, this process is understood through molecular bonding, valence electrons, and quantum fields. But these descriptions only account for the efficient and material causes.

The question remains: "Why does this specific combination yield water—a cool, clear, life-sustaining liquid—rather than fire—a destructive process of oxidation?"

Scientifically, we know that water forms when two hydrogen atoms bond with one oxygen atom due to electron configuration and bonding rules: Oxygen has 6 valence electrons and wants 8, while each hydrogen has 1 and wants 2. The sharing of electrons in covalent bonds leads to a stable H_2O molecule.

Energetics: The reaction between H_2 and O_2 is exothermic — it releases energy, which can produce fire (a combustion reaction), but the product is not fire itself, it's water.

Molecular stability: H_2O is a stable configuration based on quantum mechanical laws — the orbitals, bond angles (~104.5°), and resulting dipole moment give rise to water's unique properties. But why does this combination yield that substance? Why not something else? Evolutionary biology provides no explanation at all nor can it do so within its own discipline. Woefully inadequate as an Inference to a Best Explaation (IBE) when it comes to directedness intrinsic to being itself: intrinsic to nature. This is where essence or form enters:

In classical terms, water has a form (or essence) that determines what it is. It's not just a heap of properties; its nature makes it what it is when certain matter (hydrogen and oxygen) is combined in a certain way. That form determines its intelligibility — why it is not fire, or gas, or something else.

Fire, on the other hand, is a process, typically oxidation, not a substance. So the confusion sometimes arises from the fire released during combustion (when H_2 and O_2 react) — which is a stage in the transition to water, not the essence of the product. Understanding the form or nature of substances and how they differ from a process illustrates how metaphysical form explains why one thing is not another.

- Science tells us how it happens — through bonding, energy states, quantum chemistry.

- Metaphysics can ask why these laws are such that they produce water and not something else — looking for a deeper explanation in terms of essence, potentiality, and final causality.

Such questions invite a metaphysical answer:

- Formal cause: The form of water determines its identity; it is not merely a bundle of properties, but a structured unity with potentialities distinct from other combinations.

- Final cause: The hydrogen and oxygen, in the proper context, are ordered toward producing water; their interaction aims at a stable configuration—water—which fulfills certain ends in the natural order (hydration, solvent properties, thermoregulation). No intentionality is involved, but directionality is part of the nature and structured unity that determines the ordering toward producing water fulfilling a natural ends such as hydration, solvent properties, and thermoregulation.

Thus, final causality explains "why this reaction results in this product" rather than another.

17.4 Additional Illustration: Genetic Adaptation Without Evolutionary Teleology

Consider the phenomenon of gene regulation in bacteria, such as the lac operon in "E. coli", which enables the organism to metabolize lactose only when it is present in the environment. From a strictly mechanistic view, the mechanics of molecular interactions and feedback loops enables the organism to metabolize lactose only when it is present in the environment. However, from a classical metaphysical perspective:

- Formal cause: The genetic structure of "E. coli" has a particular form that includes the capacity for such regulatory behavior.

- Final cause: The behavior of the lac operon is directed toward the metabolic end of energy extraction from available sugars. This intrinsic directedness toward sustaining life is intelligible only if we posit a final cause inscribed in the organism's nature.

This illustration avoids fear of evolutionary teleology and focuses on the actual structure and functional directedness present in the organism's natural processes and nature. This illustration can be applied more generally to various biological processes where regularity and directedness are indicated to explain the intrinsic directedness that exists by virtue of the nature, essence, or ordered structure that exist in things. Of course, the metaphysics behind why this is the case was discussed in detail in the previous chapter and in the appendix. The reason it is intelligible and coherent is because of that metaphysical framework. Intrinsic natural directedness is intrinsic to being itself.

17.5 Formal and Final Causality: The Interdependence

Final causes are always instantiated through formal causes. The end (*telos*) is inscribed in the form (*essence*) of the thing. Water has a specific nature (*formal cause*) that grounds its behavior and directedness (*final cause*). Hydrogen and oxygen are not merely reacting randomly; their potentials are oriented toward producing a stable, structured outcome—"water".

Conversely, the final cause makes the formal cause intelligible: the form is not an inert blueprint but an intrinsic principle of motion and development ordered toward specific ends.

17.6 Scientific Reductionism and the Loss of Intelligibility

Modern science often describes phenomena without recourse to finality. But when it does so, it risks treating natural regularities as brute facts. Yet scientific language is full of finalistic terms—"functions," "adaptive purpose," "homeostasis," etc.—that betray an implicit teleology.

Classical metaphysics reclaims these intuitions and grounds them in a robust account of being. Final causes are not metaphysical extras, but the very grammar of nature's intelligibility. The proposed reformulation proposes a return to the intrinsic teleological structure found throughout nature. Philosophy of Science has provided a disservice to the scientific community by its wholesale abandonment of directedness in nature; directedness caused by the intrinsic natural structures found in nature itself.

Final causality is essential for a full account of nature. The example of water formation illustrates that even basic chemical interactions presuppose directionality and structured outcomes that cannot be fully explained by material and efficient causes alone. While not all natural events are directional, the widespread applicability of final causality in rendering the natural world intelligible justifies its metaphysical inclusion. Classical philosophy offers a framework where science and metaphysics are not at odds but complementary paths to understanding.

As part of the reformulation discussed in the previous book *The First Cause* and in this text *Being and Necessity*, the next chapter will propose a multimodal logic based in classical metaphysics. It will provide a means for science to discuss with rigor metaphysical facets of their own disciplines including cosmology, astrophysics, and theoretical physics beyond simply appealing to mathematical models and problematic inferences to a best explanation (IBEs). The final chapter will present an argument against naturalism being the ultimate truth bearer and ground for empirical science. As this text has demonstrated, there are many reasons why a methodological naturalism or metaphysical naturalism is insufficient and lacks the necessity required to ground empirical science. Methodological naturalism, instead of freeing science, it limits science. But science must be grounded in a rigorous classical metaphysics to move forward and to avoid the disparity of incoherence. And that is the role of the reformulation described, moving beyond the constraints imposed by *a posteriori* rigid designators and possible worlds, Kantian transcendental philosophy, and Hume's skeptical empiricism.

Follow on research might entail a detailed analysis of Kripke's possible world modal logic compared to the multimodal logic grounded in classical metaphysics. Another area of research might be to take final causality and demonstrating where it applies throughout nature using rigorous modal analytic logic: discussing both the structure that makes it possible and the intrinsic directedness in nature because of that nature, essence, or structure.

18.

EPISTEMIC ASYMMETRY AND THE METAPHYSICAL GROUND OF EXPLANATION

This chapter critiques the epistemological asymmetry in contemporary discourse surrounding the existence of God, specifically the presumption that naturalism serves as the default explanatory framework in science and philosophy. We argue that this assumption results in an unjustified double standard: theistic claims must conform to naturalistic criteria to be taken seriously, while naturalistic claims are exempt from reciprocal scrutiny. We first demonstrate that naturalism, as a metaphysical thesis, is not neutral and must be subject to the same epistemic standards it imposes. We then construct two formal arguments. The first uses quantified modal logic to argue that, if physical reality is contingent, it cannot serve as the ground of its own intelligibility. The second deploys epistemic logic to show that privileging naturalism in evaluating God's existence creates an unjustified asymmetry in standards of justification. The goal is to reframe debates over theism and science by exposing and correcting this epistemological bias.

Contemporary discourse often assumes that naturalism is the default or neutral position in discussions concerning the existence of God. Consequently, any theistic argument is expected to satisfy naturalistic standards, whereas arguments for naturalism or against God are rarely subjected to equivalent scrutiny. This chapter aims to expose and correct this asymmetry by arguing that naturalism is itself a metaphysical thesis and must be evaluated as such.

Naturalism, broadly defined, holds that reality is exhausted by the physical and that all phenomena can, in principle, be explained by natural causes. While methodological naturalism has a pragmatic role in science, metaphysical naturalism is a substantive claim about what exists. Treating it as a default assumption in scientific, metaphysical, or theological debates is question-begging.

The asymmetry lies in demanding that arguments for God's existence be constructed and evaluated within the constraints of naturalistic metaphysics, while naturalistic accounts are not expected to justify themselves in light of theistic metaphysics. This results in a self-insulating framework that disallows metaphysical rivals from even being considered on equal terms.

18.1 Modal Argument Against Naturalism as Ultimate Ground

Let: "N" denote the natural world, and "G" denote God as a necessary, self-subsistent being (argued in previous chapters).

Premises:

1. $\exists x \, (Cx \land \exists y \, Eyx)$ [There exists a contingent entity x that depends for its existence on another entity y.]
2. $\exists x \, Cx$ [There exists at least one contingent being.]
3. $\forall x \, (Cx \rightarrow \exists y \, Eyx)$ [All contingent things depend for their existence on another entity.]
4. $\forall x \, (Cx \rightarrow \neg Exx)$ [Nothing contingent is self-explanatory.]
5. $\forall x \, (Cx \rightarrow \exists y \, (Eyx \land \neg yCx))$ [All contingent things ultimately depend on a necessary being.]
6. $\exists x \, \neg Cx$ [There exists a necessary being (God).]

Conclusion:

7. $\exists x \, (Gx \land \forall y \, (Cy \rightarrow Eyx))$ [There exists a necessary being upon which all contingent beings ultimately depend.]

Unlike methodological or metaphysical naturalism, classical metaphysics can demonstrate dependency of all contingent being upon a necessary being for their existence. Privileging any form of naturalism without being able to demonstrate dependency of contingent being upon contingent being for its existence is asymmetrical and unjustified.

18.2 Epistemic Argument Against Asymmetric Standards

Let:

- Kp = "p is known"
- Bp = "p is believed"
- Jp = "p is justified"
- Dp = "p is treated as a default hypothesis"
- N = "naturalism is true"
- T = "theism is true"

Premises:

1. $KN \rightarrow JN$ [If naturalism is known, it must be justified.]

2. $KT \rightarrow JT$ [If theism is known, it must be justified.]

3. $DN \wedge \neg JN$ [Naturalism is treated as default without being justified.]

4. $\neg DT \wedge JT$ [Theism is not treated as default but is required to be justified.]

5. $Jp \rightarrow (Kp \vee Bp)$ [Justification is necessary for knowledge or rational belief.]

Conclusion:

6. $\exists p \; [(Dp \wedge \neg Jp) \wedge (\neg Dp \wedge Jq)] \rightarrow$ Asymmetry(p, q)

7. Asymmetry(N, T)

8. Therefore, the burden of justification is unfairly assigned, favoring naturalism without argument.

If theistic claims must satisfy rigorous standards of justification, so must naturalistic ones. Philosophy of science and metaphysics must acknowledge that naturalism is a metaphysical stance, not a neutral baseline. This recalibration is necessary for fair debate and robust inquiry. It is also essential for empirical science to do proper research beyond methodological naturalism because it leads to confirmation bias and a failure to consider other possible IBEs.

The privileging of naturalism as the epistemic ground for evaluating a scientific hypothesis or theism is unjustified and masks a metaphysical commitment. Correcting this asymmetry opens space for more honest and fruitful dialogue between science, metaphysics, and theology.

The claim "you cannot prove God exists from physics" does not entail "God does not exist," unless one assumes that physics exhausts all legitimate forms of inquiry—a claim that is itself **metaphysical**, not physical.

Likewise, the claim, "from physics you cannot claim God created the universe, therefore God did not create the universe." Since physics cannot prove why there is anything at all or why the universe is intelligible, demanding such proof from theism but not from naturalism is an **unjustified asymmetry in epistemic standards**. Our argument shows that theism offers a coherent metaphysical explanation for contingency that naturalism cannot supply without borrowing metaphysical assumptions it cannot justify.

18.3 A Socratic Dialogue

Lucretius: so is there a problem with me asking a proof of God's existence or a proof god made the universe from physics?

Socrates: not at all. But there are a few things one should realize about demanding proof from physics.

Lucretius: well science is about nature and there really is nothing beyond nature. So it should be science that determines if god exists or created the universe, and provide proof of god's existence and proof god made the universe.

Socrates: it seems like this is a double standard, isn't it?

Lucretius: how so?

Socrates: Such a position reveals a double standard in how naturalism is often treated: it's taken as normative or default in scientific discourse, such that any evidence for God must be filtered through naturalistic assumptions, but evidence against God need not undergo the same scrutiny.

Lucretius: I don't understand.

Socrates: do you agree that in science, methodological naturalism is the principle that we look for natural causes for natural phenomena. It's pragmatic and often productive.

Lucretius: Yes, of course. Science must assume a closed system to avoid crazy stuff. For example, suppose I take two hydrogen atoms and combine them with one oxygen. I always expect water. That's the chemical composition of water. If I don't assume naturalism and a closed system, if I discover that H3O is water on some yet unknown planet where it might be stable, I might attribute this to some miracle rather than a valid chemical compound for water in some unusual stable case.

Socrates: Fair enough and absolutely correct. But methodological naturalism is the principle that we look for natural causes for natural phenomena, and that is sometimes conflated with metaphysical naturalism—the claim that only natural things exist. That is the step you are making when asking for physics to prove God created the universe or by stating there must be a closed system at all.

Lucretius: so, are you saying, if one starts by assuming metaphysical naturalism, then by definition, no argument from physics could prove god, because god isn't part of the physical/natural order?

Socrates: You obviously know this and that's why you're asking the question. You don't expect an answer, nor do you want an answer in my opinion, prove me wrong.

Lucretius: I have repeatedly asked you to prove god from physics. If you can't, obviously god doesn't exist.

"Only physical causes are real; therefore, God (if not physical) isn't real." That should be obvious to anyone.

Socrates: This is indeed the assumption you're making. It's a circular argument based on metaphysical naturalism.

Lucretius: No, it's a reasonable question. You claim that god created the universe and I would like to see the physics research for that claim.

Socrates: That's like asking how much the "motive" weighs in a murder trial.

Lucretius: physics deals with empirical, measurable phenomena—things that can be observed, modeled, and predicted within spacetime. So god, if god is real and exists, should be observable from brute facts. Or, modeled like quantum mechanics or the singularity. Or, predicted using Bayesian logic like autonomous cars predicting a pet running across the road, or predicting behavior in relativity, or quantum mechanics. There should be a clear way to see god's existence or that god created the universe, if god indeed exists and created the universe.

Socrates: indeed one can, a cause is known from its effects. But naturalism says, all effects have a natural cause and cannot be otherwise. So even if an effect exists from a divine cause, it must necessarily be understood as a natural cause. Not caused by god using natural laws, or a deity intervening in the natural world. For example, if one investigated the probability of those who become healthy from cancer after prayer compared to those who don't. If more get well after prayer than without, one could statistically measure this. One could even determine the conditions by which this takes place. But if one assumes that the cause must necessarily be related to emotional response or natural laws, then any divine cause is simply ignored as impossible and absurd.

Lucretius: so prove god from physics. If god's effects can be seen, you still don't know it is god. It's like evolution, it's all natural effects and god did not create natural selection. They are natural effects from a natural cause. You obviously can't prove god if all effects have natural causes.

Socrates: god, in classical theism, is understood as immaterial, non-contingent, timeless, and beyond physical reality. So, asking physics to prove God's existence is a category mistake: using a tool designed for studying the material world to detect or prove something immaterial and transcendent. You're making a category mistake like asking one to physically weigh motives in a murder case. Or using a metal detector to find plastic cups.

Lucretius: But if god is immaterial, timeless, and beyond physical reality, this god must be imaginary. Like ghosts and goblins. Nothing physical about god.

Socrates: since you believe this is the case, prove god does not exist using physics. If god is just imaginary, then prove god doesn't exist.

Lucretius: the burden of proof is on you. You have to prove god exists using physics.

Socrates: your expecting naturalism to be the ground by which we prove god, but not the ground by which we disprove god's existence?

Lucretius; what do you mean?

Socrates: physics relies on testable, falsifiable models. But god is not a scientific hypothesis in the traditional sense. Do you agree?

Lucretius: Obviously, science is inherently methodologically naturalistic—it studies causes within nature, not beyond it. That's what I have been saying.

Socrates: that's right. So, any appeal to god within physics is usually a philosophical interpretation, not a scientific conclusion. Physics might suggest certain features of the universe (e.g., fine-tuning, the Big Bang, order, intelligibility) that point to god's existence as effects. This causality is called divine action, but interpretations of these facts vary.

Lucretius: some say these imply a designer. Like those who believe in creation or intelligent design. But these could just be effects of the multiverse or just brute facts about nature without a divine cause.

Socrates: and that's really the problem, physics provides data, but metaphysical conclusions go beyond what physics alone can deliver. And that is why arguments from metaphysics are valid and sound arguments to address issues that go beyond what physics alone can deliver.

Lucretius: so, are you saying I'm asking the wrong question?

Socrates: the problem isn't in asking the question—but in expecting physics, by itself, to answer a metaphysical question. If I ask you to prove god does not exist using physics, you will say that's not your burden of proof. Instead, you will say that it's my burden of proof because I said god created the universe. But what you are really saying is that you can't use physics to prove god does not exist. So you say instead, prove god exists from physics shifting unfairly the burden of proof on me. Naturalism is privileged. It has no burden of proof to prove god does not exist, it's burden of proof is to show everything is caused by natural laws that science can observe and predict.

Lucretius: what do you mean by saying that I shouldn't use physics alone and natural causes by themselves? And what do you mean, naturalism is privileged?

Socrates: to approach God's existence responsibly using physics, one must engage in philosophical reasoning informed by physical insights, not treat god as a physical hypothesis.

Lucretius: so, are you saying that's my problem, the problem is with my starting point? I don't understand the problem, nor the solution.

Socrates: often, when a theist uses physical evidence (e.g., cosmic fine-tuning), they're told:

"That's not science. That's metaphysics."

But when a naturalist claims, for example, that quantum mechanics undermines free will or that cosmology makes God unnecessary, those claims are often taken as scientific or even conclusive—without the same demand for philosophical rigor. This is where metaphysical naturalism is privileged.

Lucretius: are you saying, that's an inconsistent application of epistemic standards and a hypocritical double standard?

Socrates: naturalism is sometimes treated as the "neutral ground", and theism as a "positive hypothesis" requiring extraordinary evidence. But that ignores the fact that naturalism is itself a metaphysical position. It carries assumptions about causality, ontology, and explanation. And the same demand of extraordinary evidence is not expected or demanded by those appealing to it to justify their claims, commitments, and presuppositional biases.

Lucretius: so, you are saying, if we're going to be fair, we must compare worldviews on equal philosophical footing—not treat one as the default and the other as an intruder? Rejecting one, while embracing and affirming the other?

Socrates: it is unjustified to demand that theism meet the standards of naturalism, if naturalism itself is not held to those same standards when arguing against theism. So, you cannot expect theists to prove god exists or created the universe from physics, if you claim all evidence of god's existence is simply caused by natural laws or brute facts. Nor can you expect a heavier burden of proof to prove god's existence from physics, than you're willing to expect from those who deny god's existence.

Lucretius: But you said god created the universe, you can't prove god created the universe from physics, so god didn't create the universe obviously.

Socrates: naturalism itself cannot prove from physics why anything exists at all, or why the universe is intelligible, ordered, coherent, has regularity following natural laws, and why it is contingent (i.e., why it comes to be and passes away). If naturalism is treated as the default without being justified, and theism is dismissed for failing to meet the unjustified default's criteria, then we have epistemic asymmetry. In other words, imagine two people in a debate. One is allowed to speak without evidence because they're considered "neutral," but the other must present flawless evidence just to be heard. That's asymmetry—an unfair standard that favors one side without justification.

Lucretius: the burden of proof is on you not on me because you said god created the universe.

Socrates: this is unfair and irrational, because both views are metaphysical positions—neither can be proven by physics alone, and both make claims that go beyond empirical data.

Lucretius: unless you can prove god emperically, belief in god is irrational.

Socrdates: you don't apply the same standard to naturalism itself— why assume the universe exists without a cause?

Lucretius: I don't say the universe came into existence without a cause, but it was a natural cause like dark matter.

Socrates: The use of metaphysics is a legitimate form of philosophical inquiry, especially for ultimate questions where physics is limited. God, as traditionally conceived, is not a physical entity within the universe, but the necessary ground of being or ultimate cause of all contingent reality. Expecting empirical science to directly detect a metaphysical cause is like expecting a metal detector to find abstract objects—it's a misuse of tools. Investigating the ultimate cause that created the universe is not a scientific question, it is a question for the discipline of metaphysics. What can be observed from the senses and empirical science are the effects, and from the effects one can abstract the universality and necessity of a First Cause-- God is the very act of being itself, subsisting through itself, eternal: *Deus est ipsum esse subsistens per se aeternum.*

Lucretius (gesturing with his hands): Socrates, I hold that all is atoms and void. If water were changed into wine, it must be by rearrangement of matter, by hidden motions we do not yet know and measturable.

Socrates (smiling): Then let us examine, Lucretius, how such a change might be measured. Tell me: when wine is made naturally, what residues does it leave?

Lucretius: Yeast, bubbles of air, traces of heat, sediments — the marks of fermentation. If wine appeared without these, it would lack its natural history.

Socrates: Excellent. Then the *first sign* of direct transformation would be the **absence of fermentation residues.**

Lucretius (nodding): Yes, the absence of what should be there if time had passed. Very well, I grant this.

Socrates: Next: when nature works, is her handiwork uniform or always the same? Or does it show variation — off flavors, uneven mixtures?

Lucretius: She varies. No two wines are identical, for chance rules the swirl of atoms.

Socrates: Then if wine appeared of improbable perfection — balanced beyond chance — would this not be a *second sign* of something above nature was involved?

Lucretius (slowly): Yes. **Improbably perfect composition** would indeed mark it as something beyond ordinary fermentation.

Socrates: And third: when natural causes work, they pass through stages, do they not? The grape grows, the yeast eats sugar, bubbles rise.

Lucretius: Yes, every process is a chain of motions, step after step.

Socrates: But if water were suddenly wine, without intermediate steps, would that not be a *third sign* of an **instantaneous transition without physical intermediates**?

Lucretius: It would. Such a leap is alien to the dance of atoms.

Socrates (pressing gently): Now, Lucretius, science can register these signs: no residues, improbable perfection, abrupt transition. And what shall we call such a gap in natural causation?

Lucretius: A causal gap. The effect exists, but no sufficient natural cause appears.

Socrates (eyes narrowing): And can potential being — mere potency without act — ever bring itself to be?

Lucretius (sighing): No, potency needs act. That much I cannot deny.

Socrates: Then if the cause cannot be within the chain of potencies, it must be transcendent. And what is without potency, pure actuality?

Lucretius (quietly): Pure act. The sustaining ground.

Socrates: Which the wise Thomas who will join us later calls *ipsum esse subsistens* — Being itself, subsisting. So from effect to cause, from science to metaphysics, we must conclude: water-to-wine points not to atoms alone, but to Pure Act."

Lucretius (softly): Socrates, you have trapped me again. If such wine exists, it is not the child of atoms, but of Being itself. Strange... that a cup of wine could be a window into the ground of all.

Socrates: Not strange, my friend, but fitting. For wine gladdens the heart, and truth gladdens the soul.

Lucretius: But Socrates you have played a slight of hand, you claimed that the cause can be inferred from its effect, and it could be scientifically measured. Where have you measured the effect? If we can observe the effects of God turning water into wine, then can we not scientifically measure those effects?

Socrates: Yes, from the effects the cause can be inferred. But one must be willing to take the evidence on its own terms and not demand a naturalistic closed system explanation that is itself impossible.

Lucretius: What do you mean by saying, one must be willing to take the evidence on its own terms and not expect a natural explanation that is itself impossible?

Socrates: Take for example water turning into wine. If H_2O were immediately changed into wine, one would expect:

- Quantum-level nuclear transmutations (oxygen into carbon, etc.),
- Electron orbital reconfigurations,
- Spectral/energy anomalies,
- Thermodynamic violations (no explosion despite massive energy differentials),
- Fine-tuned molecular assembly **of sugars, ethanol, acids, and pigments.**

Lucretius my friend, all these are measurable at the time the event occurs. But in purely natural terms this event is impossible without catastrophic radiation. If it occurs without such fallout, the only coherent account is a **non-natural cause imposing the form of wine directly**, skipping natural intermediate steps.

Lucretius: So, Socrates you are suggesting a **natural process** would destroy everything (nuclear fireball).

Socrates: That's correct Lucretius, if water "becomes wine" without these destructive effects, the **essence (form)** of wine must be **imposed directly** by an agent with causal power at the level of *esse* (existence itself), not by rearrangement of material causes. This would be a **substantial change by divine causality**, bypassing intermediate natural efficient causes. Therefore, from the effect one observes the cause.

Lucretius: But Socrates, from methodological naturalism and a closed naturalistic system of material causes, one is forced to deny the possibility of such divine action. And this is the problem with methodological naturalism, it must necessarily deny divine action to make room for its own metaphysical naturalism.

Socrates: Lucretius, that is no longer science, that is a metaphysical commitment. Therefore, to avoid these implicit commitments, one must be willing to take the evidence on its own terms and not expect a natural explanation that is itself impossible. If taken from a purely closed naturalistic explanation, miracles are impossible. Methodological naturalism can only see the fireball unless it's willing to accept the possibility of divine action.[30]

---

30 If water were *instantly* transformed into wine, several things at the quantum/atomic level would stand out and would be observable.

Nuclear Transmutation (Impossible in Chemistry Alone)

- Water contains only hydrogen and oxygen nuclei.
- Wine requires carbon (C), additional hydrogens (H), and complex molecular structures.
- Therefore, **new nuclei must appear** (oxygen must transmute into carbon and other isotopes).
- This would involve **nuclear reactions**, releasing or absorbing **huge amounts of energy** (gamma radiation, neutrinos, particle cascades).

Quantum Jumps in Electron Configurations

- Creation of C–C, C–H, and C–O bonds requires electrons to reorganize into orbitals characteristic of organic molecules.
- One would observe sudden spectral line emissions/absorptions as electrons settled into new orbitals.

Coherence/Decoherence Spike

- An instantaneous, large-scale rearrangement of trillions of trillions of particles would briefly show **non-local quantum correlations** (essentially a *discontinuity* in the wavefunction description if no

Lucretius (brightens enthusiastically): let me press the difficulty. If water were instantly changed into wine, physically that would require nuclear transmutation — oxygen atoms splitting into carbon, reconfiguring nuclei, assembling sugars and ethanol. This would release catastrophic gamma radiation, perhaps a fireball consuming the room. By physics, that is the only conceivable route. So why did the Gospel account record Jesus changing water into wine instead of a massive explosion?

Socrates (smiling): You are correct. If water were left to nature, yes — to pass from H_2O to wine would involve potencies realized by violent nuclear change. That would yield not wine, but destruction. But instead, God acts not as one cause among others, but as *ipsum esse subsistens* — Being itself. He does not hammer matter into shape by force, but bestows existence and form according to the divine ideas. The idea of wine in the divine intellect is not 'gamma rays' but 'a potable liquid of water, ethanol, sugars, acids.' To impose this essence directly is not repugnant but fitting. The miracle bypasses the intermediate potencies that in nature would cause ruin.

mediating process occurred).

- This would look like a catastrophic decoherence event—all previous correlations of water molecules gone, replaced by correlations of organic chemistry.

Thermodynamic Anomalies

- Nuclear-to-chemical scale energy mismatch means the local environment should either explode (if energy released) or freeze/collapse (if energy absorbed).
- If neither occurs, one would infer some non-natural agent "fine-tuned" the quantum transitions to bypass normal conservation expectations.

If we try to imagine a *stepwise naturalized version* of water-to-wine:

Nuclear Reconfiguration

- Some oxygen nuclei split into carbon (via e.g. O-16 → C-12 + He-4 type reactions).
- Excess hydrogen atoms combine with carbons to form hydrocarbons and alcohol groups.

Molecular Assembly

- Atoms organize into ethanol, sugars, acids, pigments through precise covalent bonding.
- This requires extreme control over quantum bonding probabilities (far beyond natural stochastic chemistry).

Solution Structuring

- These complex molecules dissolve in the remaining water to yield the right concentrations.
- For wine's properties, ratios must be finely tuned (ethanol ~12%, sugar/acid balance for taste, tannins for mouthfeel).

Stabilization

- The mixture must settle into thermodynamic equilibrium without releasing catastrophic energy.
- This would require suspension or override of conservation laws (energy, baryon number, etc.) at the local scale.

Lucretius (inquisitively): So you are saying the destructive path is real if matter alone is at work — but God does not use that path at all?

Socrates: Precisely. What is possible to God is not what is contradictory in essence. A fireball that is wine is repugnant to being *qua* being; it is no wine at all. Therefore, God cannot produce such a pseudo-object. Instead, He grants existence directly to what accords with the exemplar idea — true wine. Thus there would be no blast, no radiation, but the quiet presence of that which was always in the divine intellect: wine itself.

Lucretius: Then miracles are not violations of nature, but the First Cause giving effect without destructive intermediates?

Socrates: Correct. God does not contradict being. He gives it. The fireball belongs to potency mismeasured as act; the wine belongs to Being itself, directly bestowed. **God does not actualize contradictory composites (wine-as-fireball), which are metaphysically repugnant.** Divine causality bypasses destructive intermediate potencies, directly imposing the form and being of wine according to its divine idea. Hence a miracle of turning water to wine is intelligible as *true wine*, not radiation or a destructive fireball.

Lucretius: But how can such an effect be measured by science?

Socrates: Let's compare *tempus compressum* and *ex nihilo* creation. If God bent time and space compressing the fermentation process, how could this be measured?

Lucretius: evidence would look like a full fermentation "squeezed" into no time — all chemical/entropic traces of history present, just without the actual timeline.

Socrates: Indeed. Yet suppose time itself were not constant. Suppose the process of fermentation could be compressed—not bypassed, but accelerated—by altering the very fabric through which change occurs. Would that not preserve your beloved naturalism while Lucretius:

Lucretius: We are speaking of compressing time, but that is no ordinary notion. Time flows as the measure of motion. To compress it would be to hasten all motion, all change. But how? allowing for divine intervention?

Socrates: let us consider the teachings of a modern sage—Einstein. He teaches that time is not absolute, but bends with gravity. In regions of weaker gravity, time flows more swiftly. Processes unfold more rapidly. If the wine-making process were placed in such a region, the atoms could dance their fermentative ballet in what appears to us as an instant.

Lucretius: So you propose that the miracle was not a violation of nature, but a divine manipulation of spacetime? That the yeast did its work, the sugars fermented, and the wine emerged—because time itself had been compressed?

Socrates: Precisely. Not a suspension of causality, but its fulfillment in a higher mode. The material, formal, efficient, and final causes all remained intact. Only the temporal medium was altered. The divine did not negate nature—it perfected it.

Lucretius: fascinating. Then the miracle becomes a revelation of deeper laws, not their contradiction. The divine will, by altering gravity, accelerated time locally. The wine was not conjured, but actualized.

Socrates: and this, my friend, aligns with the medieval notion of *tempus compressum*—a mode of time not successive, but unified. In angelic cognition, truths are grasped all at once. In divine action, processes may unfold in compressed form. The miracle at Cana, then, is a glimpse of eternity touching time.

Lucretius: you have woven metaphysics and physics into a single tapestry. You have discovered the **theory of everything**. I concede: if such a compression of time is possible, then the miracle is not irrational. It is sublime.

Socrates: you say correctly, the theory of everything. The water becomes wine not by divine fiat alone, but through a compressed unfolding of natural causality, orchestrated by God within the quantum fabric of spacetime. The yeast, sugars, and molecular transformations all occur consistent with Einstein;s General Theory of relativity. Time is locally compressed via quantum gravitational manipulation. The miracle is intelligib and Quantum Gravity. It's causally coherent, and revelatory. Thus, the miracle at Cana is a perfect instantiation of *potentia ordinata*—a divine act that respects and elevates nature's intelligibility, made possible by the quantum pliability of spacetime. Let us pause a moment and consider.

Socrates: imagine that spacetime is not smooth, but made of tiny grains—like sand beneath the sea. Quantum gravity tells us that space and time are quantized, like atoms of duration and extension.

Lucretius: so time itself is built from indivisible units? Then it could be rearranged, compressed, even folded?

Socrates: precisely. Einstein showed us that gravity bends time. But quantum gravity suggests that time can be sculpted at the smallest scales. If the divine wished to ferment wine instantly, He need not break nature's laws—He could simply reshape the spacetime fabric where the water sat.

Lucretius: adivine winemaker, not by magic, but by geometry. Intriguing. But does this not violate the order of nature?

Socrates: ah, now we invoke our friend Thomas who will join us later. He taught that God has two powers: *potentia absoluta*, the power to do anything logically possible, and *potentia ordinata*, the power He chooses to use within the laws He ordained.

Lucretius: so the miracle at Cana was not a divine override, but a divine orchestration—using the deepest laws of nature?

Socrates: exactly. The wine was not conjured—it was fermented. But the process unfolded in a compressed temporal region, made possible by quantum gravitational manipulation. The Logos did not suspend causality; He accelerated it within the ordained order.

Lucretius: then miracles are not violations, but revelations. They show us that nature is more pliable, more profound, than we thought.

Socrates: and that divine action is not chaos, but clarity. The miracle at Cana is a window into a cosmos where physics and metaphysics dance together—where time can bend, atoms can sing, and wine can flow from water without contradiction.

Lucretius: Socrates, I must interrupt our celebration of compressed time. You speak of fermentation, of yeast and sugar, but you forget the vine. There were no grapes at Cana—only water. How can wine emerge without its material origin?

Socrates: You are right: fermentation presupposes grapes. But consider this—what if the divine act did not merely accelerate time, but restructured the quantum substrate of matter itself?

Lucretius: you mean the water became grapes before it became wine?

Socrates: in a sense, yes. If spacetime and matter are quantized, then the divine Logos could reconfigure the quantum fields of water into those of grape must. Not by violating nature, but by selecting a path through quantum possibility space where water's atomic structure is transmuted—legitimately—into the precursors of wine.

Lucretius: so the miracle is twofold: first, a quantum transmutation of water into grape must; second, a temporal compression of fermentation. Both within the bounds of nature, yet orchestrated by divine wisdom.

Socrates: exactly. The material cause is not bypassed—it is instantiated through quantum manipulation. The efficient cause—fermentation—is compressed through gravitational restructuring. And the final cause—joy, celebration, revelation—is fulfilled.

Lucretius: Socrates, I remain troubled. You speak of quantum reconfiguration, of water becoming grape must. But such a transmutation—if it alters atomic nuclei—would unleash energies akin to stellar fusion. Are we to imagine the wedding feast ending in a fireball? Have we not returned to our original paradox, how can natural laws be respected without distruction?

Socrates: a fair concern, Lucretius. Indeed, if the Logos were to force a change at the level of nuclear binding energies, the result would be catastrophic. But consider: the divine act need not be a brute rearrangement of protons and neutrons. Rather, it may operate through non-violent field modulation.

Lucretius: explain yourself. How can matter change without destabilizing its atomic structure?

Socrates: let us invoke the idea of field resonance. Every particle is an excitation of a quantum field. If the Logos has mastery over these fields, He need not break atoms apart—He can tune the resonances, guiding water's molecular configuration into that of grape must, not by fusion, but by field harmonization.

Lucretius: so the transformation is not a collision, but a symphony?

Socrates: precisely my friend. Think not of smashing nuclei, but of re-phasing wavefunctions. The Logos, as divine conductor, orchestrates a transition that is coherent, not explosive. The energy landscape is navigated, not ruptured.

Lucretius: then the miracle is not a violation of conservation laws, but a traversal of permitted but using quantum pathways?

Socrates: That's correct Lucretius. But what of *ex nihilo* creation? What if the Logos does not merely reconfigure what is, but calls into being what is not?

Lucretius: you speak now of a deeper mystery. Transmutation assumes substrate—fields, particles, spacetime. But ex nihilo? That is not transformation, but ontological inauguration. How can something arise from absolute nothing?

Socrates: let us be precise. "Nothing" here is not vacuum, nor quantum potentiality. It is non-being—no energy, no field, no metric. And yet, the Logos speaks, and being erupts into the void.

Lucretius: but this violates every conservation law. No symmetry, no prior state, no causal chain. Is this not irrational?

Socrates: only if we confine causality to temporal succession. But the divine act is not temporal—it is ontological causality, where the Logos is not a prior state, but the ground of possibility itself. He does not act within nature; He authors nature.

Lucretius: then creation is not a change, but a transition from non-being to being, without intermediate. No energy is released, because no energy was stored. No fireball, because there was no fuel.

Socrates: exactly. The miracle at Cana may be a transformation, but *creatio ex nihilo* is a metaphysical singularity—not a point in spacetime, but the birth of spacetime itself. It is not a rearrangement, but a donation of existence.

Lucretius: then the Logos is not a craftsman working with tools, but a composer who writes the score itself—and with it, the laws by which it may be played.

Lucretius: I suppose if God condensed the fermentation process into no time and we could trace the chemical/entropic traces, I suppose using *ex nihilo* creation that evidence would look like wine with no causal history — no yeast DNA, no CO_2 build-up, no entropy trajectory. Almost like the wine had just dropped into reality at its finished state.

Socrates: Right again Lucretius. One would analyze **gas chromatography & isotope** to check whether byproducts are consistent with real fermentation or "implanted" composition. Science would check the **radioactive clocks** (C-14, tritium, K-40) to see whether time actually passed for the matter. Then make an entropy audit whether disorder accumulated via process or appears at an end-state.

Lucretius: So, if wine appears by **time compression**, science would measure all the "signatures of history" (byproducts, entropy, isotopic clocks) but with no elapsed macroscopic time. But on the other hand, if wine appears **ex nihilo**, you would measure a "perfectly finished product" without historical residues — like matter with no past?

Socrates: Indeed.

Lucretius: then we would treat the wine as a kind of forensic artifact. Gas chromatography would reveal whether esters and aldehydes formed through enzymatic pathways or were simply present. Isotope ratios— carbon-13, oxygen-18—might betray whether the grapes ever grew.

Socrates:: exactly. And if the wine were created *ex nihilo*, those ratios might be internally consistent yet historically incoherent. No trace of photosynthesis, no microbial lineage.

Lucretius: and the radioactive clocks—C-14, tritium, potassium-40—would tell us whether time passed for the atoms. If they show no decay, then the wine is temporally newborn. If they show decay inconsistent with fermentation, then something stranger is at play.

Socrates: yes. The miracle might simulate age, but without the thermodynamic scars of process. An entropy audit would reveal whether disorder accumulated gradually or was implanted as a final state. Like a novel written with no drafts.

Lucretius: so the wine could be indistinguishable from aged wine in taste and structure, yet bear no causal genealogy. It would be a metaphysical orphan—complete, coherent, but without a past.

Socrates: Lucretius, if you were to taste the wine, what would you expect the wine to taste like?

Lucretius: ah, Socrates, you tempt me into poetry. If the wine were truly divine—whether by compressed time or *ex nihilo*—I would expect it to taste like the *telos* of fermentation itself. Not merely good wine, but wine as it ought to be. It would truly be a Divine idea of what wine ought to be.

Socrates: Oh, I wish Plato were here, so not just the product of grapes and yeast, but the ideal form of wine?

Lucretius: yes. A Platonic vintage. The wine would bear no flaws, no overtones of decay or imbalance. It would taste as if purpose had been distilled into liquid—a harmony of sugars, tannins, acids, and aromas, not by accident, but by intention.

Socrates: then you would not merely be tasting molecules—you would be tasting meaning.

Lucretius: if time was compressed and fermentation was used, it would taste like fermented wine. It would likely taste like a clean, balanced, smooth, earthy or spicy enjoyable wine with fermentation history or residue. But *ex nihilo* it would be a "perfectly finished product" without historical residues — like matter with no fermentation residue. It would be like a temperature-controlled fermentation process, avoiding oxygen and creating a fresh and fruity flavor avoiding fermentation residues. It would be of exceptional quality (*kalos*) in every way in taste and in character.

Socrrates: exactly. And that is the paradox: the wine would be **empirically perfect**, yet ontologically unsettling. For it would lack the scars of process, the fingerprints of time. It would be too good, too complete, as if reality had skipped the struggle and arrived at the end.

Socrates: and that is your answer my friend Lucretius. *Tempus compressum* is miraculous because natural causes operate in their essence, but time-order is transcended. *Ex nihilo* is miraculous because matter and form are created directly, without any natural causes or processes at all but the taste would be of exceptional quality (*kalos*) in every way in taste and character. God acts not as one cause among others, but as *ipsum esse subsistens* — Being itself. He does not hammer matter into shape by force but bestows existence and form according to the divine ideas as a composer in harmony with nature and this explains the *kalos* in character and the taste mentioned in the Gospel of John at the wedding where Jesus turned water to wine. And the effect would have been scientifically measurable in the taste and quality of the wine.

Lucretius: So let me summarize: water-to-wine, if measured scientifically, would show (1) absence of fermentation residues, (2) improbably perfect composition, and (3) instantaneous transition without physical intermediates if the wine itself was create *ex nihilo*. It would be the absence of what should be measurable that would indicate from the effect, the cause. But if the grapes were created *ex-nihilo* and went through compressed time: every particle would be an excitation of a quantum field. The resonances would be tuned, guiding water's molecular configuration into that of grape must, not by fusion, but by field harmonization. But how would this be measured Socrates?

Socrates, you speak of quantum fields and tuned resonances, of water becoming grape must not by fusion, but by harmonization?

Lucretius, but I ask again—how would this be measurable? Could science detect the divine tuning?

Socrates: a subtle question, Lucretius. If every particle is an excitation of a quantum field, then the transformation would involve reconfiguring field amplitudes, not smashing nuclei. The divine act would be coherent, not violent.

Lucretius: then the wine would be stable, its atoms intact. But could we detect the tuning? Would the field harmonization leave a signature?

Socrates: perhaps. If the transformation occurred within spacetime, even compressed, we might find anomalous coherence—a molecular structure too perfect, too low in entropy, too free of stochastic noise. The wine might lack the statistical fingerprints of natural fermentation.

Lucretius: so we would look for non-random distributions of isotopes, unusually uniform ester profiles, or quantum coherence beyond decoherence thresholds?

Socrates: exactly. The wine might exhibit quantum order that natural processes cannot sustain. Like a crystal formed without flaws, or a melody composed without dissonance.

Lucretius: and if the grape must itself were created ex nihilo, then even the quantum fields would be newly instantiated. No vacuum fluctuations, no zero-point noise inherited from prior states.

Socrates: yes. The wine would be ontologically fresh—not just temporally young, but causally unanchored. It would lack the quantum scars of history. Its field excitations would be pristine, as if reality had just begun.

Lucretius: then science might detect not what was done, but what was not done. The absence of causal residue. The silence of history.

Socrates: precisely. The miracle would be measurable not by its presence, but by its lack of process. A kind of ontological vacuum filled with purpose.

Lucretius: from these effects, science can measure that there is a causal gap. But Socrates, how do we determine the cause?

Socrates: I see, you are returning to our original discussion, couldn't there have been **a self-actualized being**, possibly within the universe itself, or in the singularity: something that has *always* existed, and that could be the cause of a singularity?

Here we will leave our companions deep in thought. The discussion will move to an auditorium to accommodate a renowned and popular guest.

19.

MODERN SCIENCE AND THE METAPHYSICAL GROUND OF EXISTENCE

What is the role and limitations of modern science. We hear every day of a new discovery that pushes God out of the picture. Philosophy, theology, and the notion of God are all antiquated ideas are they not? Or, is there more to the story than what we have been taught and persuaded to believe?

Natural sciences—physics, chemistry, biology, geology, astronomy—do not presuppose "design" in nature. Their primary task is to observe, describe, and explain regularities, patterns, and structures in the natural world. They build models, propose laws, and test hypotheses to account for phenomena. Whether those patterns are "designed" or not is not a scientific question but a philosophical or metaphysical one.

Many natural systems exhibit:

- Complexity (intricate structures like the eye or DNA),
- Functionality (organs or ecosystems work toward apparent goals),
- Efficiency (photosynthesis converts energy better than many human devices).

This "design-like" quality is undeniable—but science treats it as the result of natural processes (evolution, self-organization, physical laws), not necessarily intentional design.

Does that mean design did not occur, the answer is no. Does that mean science should entertain the possibility of ID, the answer is yes. Does that mean science will do so, the answer is no. Should they do so, the answer is science should be about discovery not presuppositional bias, not implicit bias, not confirmation bias in either direction. Science is never a neutral actor in the dialogue of discovery.

Biomimetics (or biomimicry) does indeed look to nature for solutions. Engineers and designers copy:

- Velcro from burr hooks,
- Aerodynamic shapes from birds or fish,
- Efficient adhesives from geckos.
- Photosynthetic efficiency in solar panel designs.

But note the subtlety: biomimetics learns from evolved solutions, not necessarily designed ones. Evolutionary biology explains these features as products of adaptation and selection. From a scientific standpoint, they are functional patterns shaped by natural history. But due to presuppositional bias of a closed natural system and a claim to neutrality, other alternatives are NOT entertained. It is not merely a claim to a privileged neutrality; it is an overt commitment to natural explanations as the sole source and method of scientific inquiry that is the issue.

Science: should study "what is there" and "how it works." If there was water-to-wine, it should be studied because it is there and how it occurred, and the effects should be analyzed. It is the role of metaphysics to take the results of what has been analyzed and to explain how it came about when it is not explained by natural causality as we saw with water-to-wine.

Philosophy/theology: asks "why such ordered complexity exists at all" and whether "design" is the best ultimate explanation. And it asks, why did it come about in the first place using natural causality, natural evolutionary forces, natural selection, natural process.

So yes—natural sciences do discover patterns and efficiencies in nature, and biomimetics shows how fruitful it is to copy them. But calling them designs depends on whether one means "intelligent design" (philosophical/theological claim) or "functional structures shaped by natural processes" (scientific claim). Both can be true simultaneously and that is the problem. Science and religion are both engaged in a false dichotomy. Both sides are paranoid of the other and both sides use propaganda to justify their warranted belief.

Natural science discovers highly ordered patterns and efficiencies in nature, and biomimetics is a striking example of how fruitful it can be to imitate them. Velcro from burrs, aerodynamic structures from birds, or energy capture modeled on photosynthesis all show that natural systems achieve a level of efficiency and ingenuity that human engineers often struggle to match.

But here we have to distinguish science from philosophy: Science studies how these structures arise, function, and adapt. It does not presuppose design but explains complexity through processes like physics, chemistry, and evolution. For science, the "design-like" character of nature is simply the outcome of natural processes.

Philosophy and metaphysics, however, ask the further question: Why is there such rational order in nature to be discovered in the first place? Why is the world intelligible, with stable laws and structures capable of yielding such remarkable efficiencies? This is not something empirical science explains, but it is the very condition of possibility for science.

From a classical metaphysical perspective, this intelligibility of nature is not accidental. Philosophers like Aquinas, Averoes, Avicenna, Ockham, and numerous others would argue that the pervasive order, directedness, and efficiency discovered in natural things reflects participation in the eternal reason (*ratio aeterna*) of God. So while biomimetics highlights the usefulness of nature's patterns for technology, philosophy sees in those same patterns a deeper metaphysical pointer: order in the effects implies order in the cause, and the ultimate cause cannot be less than intelligent but must be pure act — *ipsum esse subsistens*. Let's walk through an argument and formalize it.

Premise 1. Natural sciences discover highly ordered, functional, and efficient structures in nature (e.g., photosynthesis, bird flight, gecko adhesion).

Premise 2. These structures exhibit design-like features: complexity, efficiency, and directedness toward specific ends.

Premise 3. Science can explain how such features arise through natural processes (e.g., evolution, physical laws), but it does not and cannot explain why the natural world is intelligible and ordered in such a way that these features exist and can be discovered.

Premise 4. The existence of pervasive order and intelligibility in nature calls for an explanation that is not reducible to the contingent processes within nature.

Conclusion. Therefore, the intelligible order of nature ultimately points beyond itself to a transcendent source of order — what classical metaphysics identify as *ipsum esse subsistens* (subsistent being itself, pure act).

Let's turn this into a modal argument.

Modalities: □ = necessarily, ◇ = possibly.

Quantifiers: ∀x (for all x), ∃x (there exists an x).

Domains & Predicates

N(x): x is a natural system/structure (within nature).

O(x): x exhibits order/functional organization.

I(x): x is intelligible (law-governed, modelable).

C(x): x is contingent (its being/order could have been otherwise).

E(y,x): y explains x (y is a sufficient reason/cause of x's being/order).

⊑: mereological "part of" (y ⊑ U = y is within nature's contingent totality).

U: the totality (aggregate) of contingent natural realities.

G(y): y is a transcendent, non-contingent ground of natural order (what metaphysics identifies with *ipsum esse subsistens*).

Abbreviations

SR(x): x has a sufficient reason/explanation ($\exists y\, E(y,x)$).

Nec(y): y is non-contingent ($\neg C(y)$).

Formal Argument

P0. Empirical Order Found in Nature

$\exists x\, (N(x) \wedge O(x))$

P1. Order $\rightarrow$ Intelligibility (scientific modelability)

$\Box \forall x\, (O(x) \rightarrow I(x))$

P2. Intelligible Naturals are Contingent in their particular ordering

$\Box \forall x\, ((N(x) \wedge I(x)) \rightarrow C(x))$

P3. Modal PSR (Sufficient Reason for Contingents)

$\Box \forall x\, (C(x) \rightarrow \exists y\, E(y,x))$

(equivalently: $\Box \forall x\, (C(x) \rightarrow SR(x))$)

P4. Define the Contingent Totality U

$U = \Sigma\{x \mid N(x) \wedge C(x)\}$

("U" is the mereological sum/aggregate of all contingent natural realities.)

P5. No Self-Explanation of the Totality by a Proper Part

$\Box \neg \exists y\, ((y \sqsubseteq U) \wedge E(y, U))$

P6. The Totality is Contingent

$\Box (C(U))$

C. Transcendent Necessary Ground

From P3, P5, and P6:

$\Box \exists y\, (Nec(y) \wedge E(y, U))$

Definition $\rightarrow$ Theistic Conclusion

$\Box \exists y\, (G(y))$, where $G(y) := Nec(y) \wedge E(y, U) \wedge (y \notin U)$

("There necessarily exists a transcendent, non-contingent ground that explains the ordered, intelligible totality of nature.")

In simple English:

P0: Science actually finds ordered structures in nature.

P1: If something is ordered, it is (robustly) intelligible—fit for law/model description.

P2: Natural intelligibilities are contingent (they need not have been as they are).

P3: PSR (modal form): every contingent has a sufficient explanation.

P4: Gather all contingent natural realities into a single explanandum, U.

P5: No member/part of U can explain U as a whole (no total self-explanation by a proper part).

P6: The totality U is itself contingent (its being/order could have been otherwise).

C: Therefore, there must be a non-contingent (necessary), transcendent explainer of U—what classic metaphysics identifies with *ipsum esse subsistens*.

You can compress this to simply;

$\exists x \, (N(x) \wedge O(x))$

$\Box \forall x \, ((N(x) \wedge O(x)) \rightarrow C(x))$

$\Box \forall x \, (C(x) \rightarrow \exists y \, E(y,x))$

$U = \Sigma \{x \mid N(x) \wedge C(x)\} \,\, \& \,\, \Box C(U)$

$\Box \neg \exists y \, ((y \sqsubseteq U) \wedge E(y,U))$

$\therefore \Box \exists y \, (\neg C(y) \wedge E(y,U))$ (call such a y, G)

Let us now continue our Socratic Dialogue and discuss the origin of the universe. This will allow us to pull it all together.

20.

THE SINGULARITY

Characters

Lucretius — Roman poet-philosopher, materialist.
Socrates — the gadfly, master of ironic questioning.
Thomas — a classical metaphysician, calm and weighty.
Carl — a modern scientist, eloquent and poetic.

Setting

A modern university auditorium. Spotlights illuminate a long table where the four thinkers sit. The hall is packed with students, scientists, and philosophers. Murmurs fade as the moderator introduces the panel. Applause dies down. A contemplative hush fills the room.

[Lucretius rises, his voice deep and rhythmic, as if chanting poetry.]

Lucretius: Returning to our original discussion, couldn't there have been **a self-actualized being** within the universe itself, one that had *always* existed, and that could be the cause of the singularity? Why invoke gods or metaphysical mysteries? The atoms and the void are eternal. They move, combine, dissolve, and recombine. The singularity was no birth, but a contraction, the Big Bang, no creation, but a release. Nature is self-sufficient, its breathing without beginning or end.

[Soft applause. Sagan leans in, voice measured, resonant.]

Carl:

The cosmos is all that is, or ever was, or ever will be.
Perhaps an eternal quantum vacuum underlies it all,
a sea of fluctuating fields.
Or the timeless laws of physics,
governing without change.
If anything is self-actualized,
it is the cosmos itself.
We need not look beyond.

[The audience nods. A ripple of approval moves through the hall.]
[Socrates stands slowly, smiling wryly, eyes twinkling.]

Socrates:

My dear Lucretius and Carl,
let me ask:
what is it to be self-actualized?
Surely it is to lack all unrealized potential.
But your atoms collide and dissolve,
your vacuum fluctuates,
your laws bend with spacetime.
Are these not marks of *potency*?
That which can change cannot be pure act.
Tell me, then: how can the mutable
be self-actualized?

Lucretius: Then the quantum vacuum — always bubbling with energy, fluctuations eternal.

Socrates: But fluctuation is the very sign of potency: unrealized states becoming realized stochastically. Does this not presuppose a deeper actuality?

Thomas: Indeed. The vacuum is potential, not pure actuality. It cannot be self-actualized.

[The audience murmurs. Thomas rises, deliberate and steady.]

Thomas:

To be self-actualized is to be uncaused,
incapable of not existing.
It is infinite actuality,
with no potency unrealized.
It is the sustaining ground of all else,
from which every finite being derives existence.

Carl: Surely spacetime, fundamental and eternal, could be self-actualized.

Socrates: But spacetime curves, warps, expands, contracts. To bend is to have potency. That which has potency is not pure act.

Thomas: Therefore, spacetime cannot be self-actualized.

Lucretius: Very well — energy, which cannot be created or destroyed. Eternal in itself.

Socrates: But energy only exists *in something* — in particles, in fields. It shifts forms constantly. It is not self-subsisting act, but a measure of potency realized in different ways.

Thomas: Therefore energy is contingent, not necessary.

Lucretius (defiant, voice rising):
But Thomas,
why not accept eternal matter?
Why raise Being above the cosmos?

Socrates (ironically, to the audience):
Because, Lucretius,
what is full of potential cannot explain itself.
Eternal atoms are like a man
who must always eat to live —
he is never self-fed.

Carl (gesturing toward Thomas, sharp):
Then perhaps the laws of physics are eternal,
unchanging,
necessary.
Why invoke a God when equations suffice? Perhaps the laws —
timeless, unchanging, elegant equations govern all.

Thomas: But laws are not subsistent realities. They describe how beings act, but cannot exist without subjects to govern. A grammar without words, a melody without notes. They presuppose being, not supply it.

Socrates (to the audience): Then laws too fail.

Thomas (firm, solemn):
Carl, a law without being to govern is nothing.
A grammar without words.
A melody with no notes.
They presuppose being.
Thus, they cannot be self-subsisting act.

Thomas: Matter changes.
Spacetime bends.
Energy shifts form.
Laws are but descriptions —
they do not exist on their own.
All these are composites of act and potency,
contingent,
capable of not being.
They cannot be self-actualized being.

[The lights dim slightly, focusing on Socrates. He steps to the edge of the stage, speaking directly to the audience.]

Socrates:

Let us then put it plainly. All candidates fail. Each is composite, finite, conditioned. They are not self-actualized being, but always-existing potencies, flowing in and out of states.

If there were something in the universe
that was truly self-actualized,
it would not be merely one thing among other things.
It would:
- Be uncaused, unable not to exist.
- Be infinite actuality, no potency unrealized.
- Be the sustaining ground of all else, such that everything finite derives being from it.

Lucretius: so you are saying, this description fits no physical substrate.

Not matter.
Not vacuum.
Not spacetime.
Not energy.

Socrates: correct, it fits only what Thomas names
ipsum esse subsistens —
Being itself, subsisting.
[Thomas rises, his voice a low thunder.]

Therefore, You cannot have a necessary being
within the order of contingent beings.
Necessity must be transcendent,
not immanent.
Therefore the only coherent candidate
for self-actualized being
is God.
[Lucretius sits silently, gaze lowered. Sagan exhales, reflective.]

Lucretius (softly):

Then matter is not enough.

Thomas (solemn, decisive):

"This description fits no physical substrate — not spacetime, not vacuum, not law, not energy. It fits only what we call *ipsum esse subsistens* — Being itself subsisting. You cannot have a necessary being within the order of contingent beings. Necessity must be transcendent. Again, the only coherent candidate for self-actualized being is God."

Carl (after a pause, almost reluctantly):
Even if one calls it "God" in metaphor,
it is not reducible to physics.
No equation contains necessity itself.

Carl (confident):
But then why not say the singularity caused itself? A perfect density, timeless, containing the seeds of expansion — why not call it the eternal ground?

Socrates (raising a brow):
Tell me, Carl, if the singularity was 'potentially' the universe, was it already the universe in act?

Carl:
No, it was only compressed potency, waiting.

Socrates:
And can that which only 'waits' bring itself into being? Can marble carve itself into a statue?

Thomas (firm, calm):

Potency cannot move itself to act. That which is potential requires something actual to bring it forth. If the singularity was a bundle of potency — compressed density, no extension, awaiting expansion — then it cannot be its own cause. Its act must come from beyond itself not from within itself because it is potency. Just as a block of marble is potentially a sculpture, so is compressed density. It requires some action outside itself to act upon the compressed density for that compressed density to become something rather than be nothing more than compressed density awaiting expansion.

Lucretius (frowning):

Then the singularity is not self-actualized being?

Thomas:

No. The singularity is contingent, finite, composite, mutable. It is potency needing act. Only pure act can be self-actualized, and such is not a thing within the universe, but the sustaining ground of all.

Socrates (turning to the audience):

Therefore we see: potential being cannot cause itself, and so cannot cause the singularity. The cause must be pure actuality. And that, as Thomas names, is *ipsum esse subsistens*.

[Socrates smiles, spreads his hands toward the audience.]

Socrates:

So it seems we are all agreed.
The singularity, like every contingent state,
requires something more fundamental than itself.
Only Being itself, pure actuality,
is sufficient.
Not the shifting cosmos,
but the transcendent ground.

If anything in the universe were self-actualized, it would not be one thing among others, but pure actuality itself — uncaused, infinite, sustaining all. No physical substrate fits this description. Therefore, the only coherent candidate is *ipsum esse subsistens* — the God of classical theism and coherently demonstrated by classical metaphysics.

[The auditorium falls utterly silent.

Then, a slow rising applause,

building to thunderous ovation.

The four interlocutors bow their heads in mutual recognition.

The stage lights fade to black.]

[The applause fades. A spotlight shifts from the panel to a figure in the audience: a contemporary theoretical physicist in a dark blazer, holding a microphone. The hall quiets.]

Physicist (measured, thoughtful):

Friends, I am a physicist. My work is on cosmic inflation, multiverse models, and the quantum vacuum. I spend my days inside equations describing the earliest instants of the universe.

We propose inflation — a burst of exponential expansion — to explain the flatness and uniformity of the cosmos. We propose multiverses — vast ensembles of universes — to explain why our constants are tuned for life. We invoke the quantum vacuum — a seething ocean of energy — as the seedbed of space and time.

But let me be honest. Each of these is a model of **potentiality**. Inflation is a potential field, waiting to roll. A multiverse is potential universes, most of them never realized. The vacuum is fluctuations of what might be, not pure actuality.

And the singularity? It is the very image of compressed potency, finite, unstable, bursting forth into expansion. None of these, as Socrates and Thomas have shown, are *self-actualized being*.

The inference is inescapable: If there is to be any coherent explanation for why there is something rather than nothing, why inflation rolls at all, why the multiverse generates any universes, why the vacuum fluctuates, there must be a ground of pure actuality — not one potential among others, but Being itself.

Physics describes the stage, the potentials, the transitions. But metaphysics names the condition for the whole: *ipsum esse subsistens*.

Perhaps Thomas and Carl are not as far apart as they seemed. For when we peer to the bottom of the equations, we too encounter the limit of potency — and the need for act. And whether one names that 'God' or 'Being itself,' it is something physics alone cannot contain.

No feature of the universe — not spacetime, not vacuum, not laws, not energy, not the singularity itself — can be self-actualized being. Each is contingent, composite, and mutable. If something has always existed as self-actualized being, it must be uncaused, infinite actuality, and the sustaining ground of all. That description belongs only to *ipsum esse subsistens* —what we call God."

21.

BEING AND NECESSITY

What is Being and Necessity? What is self-actualized Being? **Self-actualized being** = that whose act of existence (*esse*) is identical with its essence (*quidditas*). It is pure actuality, with no unrealized potential. It therefore does not come to be, does not change, does not depend. In classical metaphysics, this is ***ipsum esse subsistens***. It necessarily exists and is the ground of necessity.

Anything less than this (a field, a vacuum, a particle, spacetime itself) is a composite of act and potency. Such things can exist a long time (even "always" in a temporal sense), but they are not self-actualized; they always in potentiality until acted upon, potential not to be. Some physicists treat spacetime as fundamental. But spacetime has geometry, curvature, and thus **potency** (it can expand, contract, warp). That means it is not pure act. Others argue for a quantum vacuum. The quantum vacuum is a sea of fluctuating potentials. But fluctuation *just is* potency being actualized stochastically. So again: it presupposes a deeper actuality.

Others argue that laws of physics or brute facts are sufficient to account for existence. Laws of physics (like gauge symmetries) look unchanging. But laws are not subsistent realities; they are descriptions of how potential actualizes. They cannot "be" without something they govern. One might say, "Energy cannot be created or destroyed, so maybe energy is the eternal, self-actualized being." But energy shifts constantly. It exists only as *in something* (fields, particles). It is not self-subsisting act, but a measure of potency actualized in different modes.

All these candidates for the cause of the universe share these limitations. They are **composite** (act/potency, form/matter, law/subject). They are **finite** (bounded, proportioned, measurable). They are **conditioned** (they depend on something else for being). Thus, they cannot be "self-actualized being." At best, they can be "always-existing potencies" that flow into and out of different states (singularity, expansion, heat death, etc.).

If there were **something in the universe** that was self-actualized, it would not be "a thing among other things" at all. It would be **uncaused** (cannot not exist), be **infinite actuality** (no potency to be unrealized), and be **the sustaining ground of all else** (everything finite derives being from it). That description just is the classical theistic God, not a feature of the contingent universe nor that of process theology. Classical metaphysics independent of theology argued that you can't have "a necessary being within the order of contingent beings" — necessity must be **transcendent** of the contingent order. So, the only coherent candidate for "self-actualized being" is *ipsum esse subsistens*, not anything physical like spacetime or energy.

This matters for the singularity because the singularity (if it existed) is not self-actualized. It is a state of **maximal potency compressed** (infinite density, no extension). A self-actualized being cannot collapse or change, so it cannot *be* the singularity. Instead, the singularity would be **caused by** the self-actualized being sustaining contingent being. That means, nothing "in the universe" could qualify as self-actualized being. Spacetime, quantum vacuum, laws, and energy are all contingent composites with potency. The only thing that could "always have existed" as self-actualized being — and thus cause and sustain the singularity — would be *ipsum esse subsistens* itself: pure act, eternal, uncaused being. A formal modal-metaphysical schema that contrasts a **contingent substrate** (like spacetime, vacuum, energy), and a **self-actualized being** (*ipsum esse subsistens*) follows again demonstrating the necessity of a **self-actualized being.**

Symbols and Definitions:

$\Box\varphi$ = necessarily true (cannot not be).

$\Diamond\varphi$ = possibly true.

C(x) = x is contingent (its essence ≠ its existence).

N(x) = x is necessary (its essence = its existence).

A(x) = x is self-actualized being (pure act).

S = the singularity (Big Bang initial state).

g = God (*ipsum esse subsistens*).

Axioms

(A1) Contingency. Anything contingent could fail to exist.

$\forall x \, (C(x) \rightarrow \Diamond\neg E(x))$

(A2) Necessity. A necessary being cannot fail to exist.

$\forall x \, (N(x) \leftrightarrow \Box E(x))$

(A3) Composition. Anything composed of parts or existing in time is contingent.

$\forall x \, (\text{Material}(x) \vee \text{Temporal}(x) \rightarrow C(x))$

(A4) Singularity. The singularity is contingent (it has potency: expansion, collapse, etc.).

$C(S)$

(A5) Pure Act. If something is pure actuality, it is necessary being.

$A(x) \rightarrow N(x)$

(A6) Ipsum Esse Subsistens. There exists pure actuality (ipsum esse subsistens).

$\exists x \, A(x)$

Demonstration against Contingent Being:

1. Suppose $\exists y \, C(y)$ always exists (e.g., spacetime, vacuum, energy).
2. Then by (A1), $\Diamond\neg E(y)$. It could fail to exist.
3. Therefore, y is not necessary.
4. Therefore, y cannot explain the singularity *as a self-actualized cause*, only as a contingent state sustained by something else.

Demonstration for Self-Actualized Being

1. Suppose $\exists x\, A(x)$. (from A6)
2. Then by (A5), $N(x)$.
3. By (A2), $\Box E(x)$: it cannot fail to exist.
4. Thus, x is the necessary ground of all contingent beings (including S).
5. Therefore, the singularity's existence ($C(S)$) requires causal participation in x.

Conclusion: $\exists S\, (C(S)) \wedge (\forall y\, (C(y) \rightarrow \neg\Box E(y))) \rightarrow \text{Requires}(\exists x\, A(x))$

Explanation: If **only contingent substrates exist**, the singularity has no ultimate explanation. (Infinite regress of contingent being.) If **a self-actualized being exists**, it necessarily exists and grounds the singularity.

- **Contingent substrates** (vacuum, spacetime, energy) may "always exist" in a temporal sense, but because they contain potency, they cannot be the cause of the singularity in the sense of self-actualization.

- **Self-actualized being (pure act)** alone can exist eternally, necessarily, and be the sustaining cause of contingent states like the singularity. It is this Being that is both Being and Necessity.

The schema shows that nothing "in the universe" can be the always-existing self-actualized being that caused the singularity. Only *ipsum esse subsistens* (pure act, God) fulfills the modal condition of necessary existence without potency, and so is the ultimate cause and ground of the singularity and all of existence. In the above dialogue Carl argued that the singularity could be the cause of itself. Why can potential being not be the First Cause? Why can the singularity not be the cause of itself?

The answer is, **potency requires act to actualize it**. For example, a block of marble has the potential to be a statue, but only an agent (the sculptor) actualizes that potential. Likewise, fluctuating quantum fields or "vacuum potentialities" require something actual to bring about concrete states. **Potency cannot reduce itself to act.** Nothing can give itself what it lacks. If the singularity was "pure potency compressed," it cannot explain its own transition into expansion (the Big Bang). **Causal Priority demands actuality.** The cause of any transition must be already actual, otherwise there is no ground to move potency to act. If the singularity began as unrealized potential, it presupposes some actuality beyond itself.

This matters for the singularity because the singularity (if real) is not pure act. It is **finite** (bounded in density and extension), **composite** (space, time, energy unified at a limit), and it is **mutable** (it collapsed, then expanded). These indicate the singularity is *potency* not in act. It cannot actualize itself because it is potency. The singularity is not self-actualized, its expansion into the universe cannot be explained by its own potentiality, and the cause must be **something already actual without potency** — i.e. pure act.

The fundamental principle is **potency cannot actualize itself. Only act can actualize potency.** Therefore, if the singularity was a state of maximal potency, it must have been caused and sustained by pure act, not by itself. Potential being cannot cause itself, and so cannot cause the singularity. The cause must be pure actuality. And that is *ipsum esse subsistens.*

Symbols and Definitions:

□/◇: metaphysical necessity/possibility (in virtue of first principles).

E(x): x exists.

Pot(x): x has unrealized potency (is changeable/composite/temporal).

Act(x): x is in act (actually able to actualize another now).

PA(x): x is *Pure Act* (actus purus; no potency).

C(y→xΔ): y causes x's transition (Δ) from potency to act.

S: the initial singularity state.

Axioms

(A1) Potency needs prior act (no self-actualization).

$\forall x \; [\text{Pot}(x) \rightarrow \neg C(x \rightarrow x\Delta)]$

(What is merely potential cannot be its own sufficient cause.)

(A2) Change requires an actualizer.

$\forall x \forall \Delta \; [\Diamond(x \text{ undergoes } \Delta) \rightarrow \exists y(\text{Act}(y) \wedge C(y \rightarrow x\Delta))]$

(A3) Material/temporal/composite ⇒ potency.

$\forall x \; [(\text{Material}(x) \vee \text{Temporal}(x) \vee \text{Composite}(x)) \rightarrow \text{Pot}(x)]$

(A4) The singularity is finite/composite/temporal.

$\text{Material}(S) \vee \text{Temporal}(S) \vee \text{Composite}(S)$

(A5) Pure Act is necessary (no potency ⇒ cannot fail to be).

$\forall x \; [\text{PA}(x) \rightarrow \Box E(x)]$

(A6) If an actualizer has no potency, it is Pure Act.

$\forall x \; [(\text{Act}(x) \wedge \neg \text{Pot}(x)) \rightarrow \text{PA}(x)]$

(A7) Anything with potency is contingent.

$\forall x \; [\text{Pot}(x) \rightarrow \Diamond \neg E(x)]$

Derivation

1. From (A4) and (A3): Pot(S).
2. From Pot(S) and (A2): $\exists y(\text{Act}(y) \wedge C(y \rightarrow S\Delta))$.
3. From (A1): $\neg C(S \rightarrow S\Delta)$. So the y in (2) is **not** S (the cause is other than S).
4. The sufficient cause of S's actualization is some y with Act(y). Two exhaustive options:

Case 4a: Pot(y).

Then by (A2) applied to y, $\exists z(Act(z) \land C(z \rightarrow y\Delta))$. Iterating this yields either (i) an infinite regress of per-se actualizers of here-and-now potency (metaphysically impossible for a sustaining series), or (ii) termination in an actualizer with no potency.

Case 4b: ¬Pot(y).
Then by (A6): PA(y).

5. Therefore, to avoid vicious regress, there must exist **y** such that PA(y).
6. From (A5): PA(y) → □E(y). So y exists necessarily.
7. A necessary actualizer with no potency is not "one thing among others" inside the contingent order but the sustaining ground of that order. Call this **ipsum esse subsistens**.

Because (i) anything with potency cannot actualize itself and requires a concurrent actualizer, (ii) the singularity has potency, and (iii) an explanatory chain of concurrent actualizers cannot regress without a first that is in no way potential, there must be a **Pure Act** that necessarily exists and sustains the singularity's being—*ipsum esse subsistens*—which is not any physical feature (spacetime, vacuum, laws, energy, quantum waves, fluctuations, possible worlds), but the transcendent ground of them all.

22.

A MULTIMODAL LOGIC FOR CLASSICAL METAPHYSICS

In classical terms, a thing is **possible** (*possibile*) iff its essence is *non-repugnant to being* (non *repugnantia ad esse*). Anything whose description **implies a contradiction** is repugnant to the *ratio entis* (the "notion of being"), so it isn't a *res* at all—it's not even the sort of thing that could be made. Hence Aquinas's famous line: what implies a contradiction "does not fall under omnipotence," because it lacks the ***ratio factibilis*** (the character of something makeable). (ST I, q.25, a.3). For example, "square circle," "a past event that did not happen," or "S is wholly moving and wholly not-moving in the same respect and time," or "for God create a rock too big for God to pick up." These don't designate possibles; they designate **nothing**. So, in classical terms, contradictory "objects" are not grounded in being and are therefore not possible—*simpliciter*.

Put schematically (using multimodal metaphysics):

- Let $\Diamond_{(CM)}\varphi$ = "φ is metaphysically possible (Classical Metaphysics)."
- Def.: $\Diamond_{(CM)}\varphi \leftrightarrow \neg$Repugnant_to_Esse($\varphi$).
- Principle of Non-Contradiction (PNC): Contradictory(φ) $\rightarrow$ Repugnant_to_Esse(φ).
- Therefore: Contradictory(φ) $\rightarrow \neg\Diamond_{(CM)}\varphi$.

Where $\Diamond_{(CM)}\varphi$ means "**In Classical Metaphysics it is possible that φ.**"

- Pronounced: "diamond sub-CM phi."
- Meaning: possibility **grounded in being** (Classical Metaphysics), not in 'possible worlds.'
- Truth-condition (essence reading):
  $\Diamond_{\langle CM \rangle}\varphi \Leftrightarrow \neg$**Repugnant_to_Esse**$(\varphi)$ — i.e., the essence described by φ contains no contradiction with the *ratio entis*.
- Equivalent proof-theoretic gloss:
  $\Diamond_{\langle CM \rangle}\varphi \Leftrightarrow$ **Con(CM** $\vdash \varphi$**)** — adding φ to the Classical Metaphysic base principles (PNC, LOI, LEM, act–potency, PSR in the sustaining sense, etc.) does **not** yield a contradiction.

Multimodal logic lets us **separate kinds of possibility/necessity** and say how they **interact**. That avoids equivocation and modal collapse—and it matches classical metaphysics far better than a single undifferentiated "$\Box/\Diamond$".

Here's what you gain:

1. Disambiguation of modalities
Use different operators for different notions:

o $\Diamond_l / \Box_l$ = logical
o $\Diamond_n / \Box_n$ = nomological (physics)
o $\Diamond_{\langle CM \rangle} / \Box_{\langle CM \rangle}$ = Classical Metaphysics (non-repugnance to being)
o $\Box^D$ = "within divine power," $\Box^E$ = essential/quidditative, F/P = temporal, K_a = epistemic, etc.

  Result: "Resurrection" can be **not nomologically possible** ($\Box_n \neg\varphi$) yet metaphysically **possible** ($\Diamond_{\langle CM \rangle}\varphi$) without contradiction.

2. Explicit bridge principles (interaction axioms)
You can state principled relationships instead of hand-waving:

o $\Box_n\varphi \rightarrow \Box_{\langle CM \rangle}\varphi$ (laws presuppose being)
o $\Box^E\varphi \rightarrow \Box_{\langle CM \rangle}\varphi$ (what holds per essence holds per being)

o $\Box^D\varphi \rightarrow \Diamond_{(CM)}\varphi$ (what divine power can do is metaphysically makeable)

o Contradictory(φ) $\rightarrow \neg\Diamond_{(CM)}\varphi$ (PNC grounds classical metaphysical impossibility)

These let you prove exactly what crosses from one mode to another—and what doesn't.

3. Prevents equivocation & collapse

Many objections trade on sliding from one modality to another (e.g., from physical impossibility to metaphysical impossibility). With distinct operators, the slide is blocked unless a bridge axiom justifies it.

4. Fine-grained axiomatics per mode

You can tailor frames/axioms to each modality (e.g., S5 for $\Box_l$, CM or S4 for $\Box_{(CM)}$, maybe something weaker for $\Box_n$ if laws vary across physically accessible states). This keeps the logic faithful to the underlying metaphysics.

5. Clean *de re* / *de dicto* management

With essential ($\Box^E$) vs accidental vs temporal operators, it's easier to say "necessarily-of-this-thing" versus "necessary-of-the-description," which is crucial for essence/existence discussions in classical metaphysics.

6. Constructive modeling of Classical Metaphysics

An example schema one can prove with the right bridges:

o (i) Contradictory(φ) $\Rightarrow \neg\Diamond_{(CM)}\varphi$ (grounded in PNC)

o (ii) $\Box_n\neg\varphi \wedge \Diamond_{(CM)}\varphi$ is consistent (miracle cases)

o (iii) $\Box^E\varphi \Rightarrow \Box_{(CM)}\varphi$ but not conversely (essence $\subseteq$ being)

o (iv) $\Box^D\varphi \Rightarrow \neg$Contradictory($\varphi$) (omnipotence doesn't range over *impossibilia*)

7. Transparent reasoning

You can formalize lines like "omnipotence concerns all things that **have** the *ratio factibilis*" as $\Box^D\varphi \rightarrow \Diamond_{(T)}\varphi$, while still validating "square circles" as outside any $\Box^D$ because they lack *ratio entis*.

If you use a **single** symbolic modal operator, you must either bake all distinctions into one semantics (inviting illicit inferences), or keep switching informal meanings mid-proof (risking equivocation). Multimodal logic gives you the **right knobs**—distinct operators and the explicit bridges—so classical metaphysics ("being grounds possibility; contradictions lack being") can be stated and proved without smuggling in assumptions.

The idea of having more than one modal operator (time, knowledge, obligation, etc.) grows out of the 1950s–70s boom in modal logic: Prior's tense logic (past/future operators), Segerberg's two-dimensional/polymodal systems, and then a full maturation in the 1990s–2000s through standard texts and handbooks. Authoritative overviews include Blackburn–de Rijke–Venema, *Modal Logic* (2001) and the *Handbook of Modal.Logic* (2006); the SEP entry "Philosophical Aspects of Multi-Modal Logic" surveys the field and applications.

Multimodal frameworks are now routine across several analytic subfields including Epistemic logic / multi-agent knowledge (distinct K_i operators per agent; interaction axioms): classic reference Fagin–Halpern–Moses–Vardi, *Reasoning About Knowledge* (1995), Temporal logic (Prior's P/F operators; later developments in the philosophy of time and agency), Deontic logic (obligation/permission with additional normative modalities; SEP overview), Dynamic logic (program/action modalities $\langle \alpha \rangle$, $[\alpha]$; widely used in philosophy of action and CS-adjacent metaphysics), and General, philosophically oriented expositions: van Benthem, *Modal Logic for Open Minds* (2010).

Gyula Klima formalizes Aquinas's semantics of being, essence, and analogy—supplying the **semantic backbone** you'd want for a Thomistic "metaphysical modality." (While not packaged as a multimodal system, it's the closest formal treatment of Aquinas's logic-of-being.)_

Analytic Thomists (e.g., **Robert Koons, Alexander Pruss, Joshua Rasmussen**) routinely use **modal logic** in arguments for necessary being, PSR, hylomorphism, etc. This is typically **one metaphysical modality** (often S5) rather than many—but nothing stops one from **refining their frameworks into a multimodal version** (e.g., separating nomological from Thomistic possibility).

Multimodal logic as a **toolkit** was developed across the modal-logic tradition (Prior, Segerberg) and systematized by the late-20th-century handbooks. It's **widely used** in analytic philosophy (epistemic, deontic, temporal, dynamic). In future work, multimodal logic will be used. The S5 and possible worlds semantics were used primarily in this text as a transition to a Classical Metaphysical multimodal framework or calculus.

A **fully worked-out "Multimodal Classical Metaphysical System"** (with explicit operator indices for classical metaphysics vs. nomological vs. logical possibility and formal **bridge axioms**) hasn't become canonical in print yet. But the ingredients are on the shelf, and **Klima's formal semantics of Aquinas** plus standard multimodal machinery make it a natural next step—one that some analytic Thomists implicitly motivate in their work. The following sets forth a Multimodal Classical Logic (MCL):

Operators:

- $\Box_{(C)}$ / $\Diamond_{(C)}$ — *Classical (metaphysical) necessity/possibility*: grounded in being (non-repugnance to *esse*).
- $\Box_l$ / $\Diamond_l$ — *Logical* necessity/possibility (consistency with logic alone).
- $\Box_n$ / $\Diamond_n$ — *Nomological* (physical-law) necessity/possibility.
- $\Box^E$ / $\Diamond^E$ — *Essential* (per-essence) necessity/possibility.
- **Dφ** — *Divine power can bring about* φ (read: "omnipotence over φ"). (Modal if you prefer: $\Box^D$ for "it is within divine power that…".)

How to read

- "$\Diamond_{(C)}\varphi$" = "It's metaphysically (Classically) possible that φ," i.e., φ's description contains no contradiction with *being*; "$\Diamond_{(C)}$ = can be, in the order of being."
- "$\Box^E\varphi$" = "φ holds per essence"; "$\Box^E$ = holds from what the thing is."
- "$\Box_n\varphi$" = "φ holds under the actual laws of nature"; "$\Box_n$ = fixed by current laws of nature."
- "Dφ" $\Rightarrow$ "φ has the *ratio factibilis* (is makeable) by God."

Two-line simple version:

- **Metaphysical possibility** ($\Diamond_{(C)}$) means: *not repugnant to being.*
- **Contradictions** (square circles, etc.) lack being → **not possible** ($\neg\Diamond_{(C)}$).

Operators:

- **C-modality**: at least **C** ($\Box_{(C)}\varphi \to \varphi$). (You may optionally add **4** if you want S4$_{(C)}$; not required.)
- **L-modality**: **S5** (standard).
- **N-modality**: **K** (keep weak; laws could vary in accessible physical states).
- **E-modality**: **C + 4** (S4^{E}) is natural for "per-essence" stability.
- **D**: take as a primitive ability predicate; pair it with bridge principles below.

Bridge principles (Classical Metaphysics):

1. **Non-contradiction blocks C-possibility**
   o ⊢ Contradictory(φ) → $\neg\Diamond_{(C)}\varphi$.
2. **Essence subsumes being**
   o $\Box^{E}\varphi \to \Box_{(C)}\varphi$.
3. **Laws presuppose being (weak direction)**
   o $\Box_n\varphi \to \Diamond_{(C)}\varphi$. *(Physical necessity ⇒ at least metaphysical possibility.)*
4. **Omnipotence ranges only over true possibles**
   o $D\varphi \to \Diamond_{(C)}\varphi$.
   o Contradictory(φ) → $\neg D\varphi$.
5. **Miracle compatibility (no collapse)**
   o $\Diamond_{(C)}\varphi \land \Box_n\neg\varphi$ is consistent. *(Metaphysically possible though nomologically impossible.)*
6. **De re / de dicto formulations (optional but useful)**
   o From $\Box^{E}(\varphi(x))$, infer $(\exists x)\Box^{E}\varphi(x)$ under standard Barcan choices listed in the appendix.

Semantic gloss for $\Diamond_{(C)}$ (metaphysical): two equivalent readings:

- **Essence/repugnance reading:**

$\Diamond C\varphi : \iff \neg Repugnant\_to\_Esse(\varphi$

("note: φ's description contains no contradiction with the *ratio entis*.")

- **Theory-relative (proof-theoretic) reading**

Let **Th_CM** represent classical first principles (LOI, PNC, LEM, act–potency, PSR in the sustaining sense, etc.). Th_CM where Th means theory and CM means classical metaphysics.

$\Diamond C\varphi :\iff Con(Th\_CM \cup \{\varphi\})$

(note: φ is consistent with classical metaphysics).

Quick contrasts:

- **Nomological** (physically) possible $\Rightarrow$ Metaphysically possible, but not conversely (e.g., resurrection: $\Diamond_{(T)}$ true, $\Diamond_n$ false).
- **Logically** consistent doesn't always guarantee Th-CM possibility if φ clashes with classical metaphysical principles (e.g., "accident without a subject").
- $\Diamond_1$ = logically possible; $\Diamond_n$ = physically possible; $\Diamond_{(C)}$ = metaphysically possible **in the classical sense**.

Example: "square circle" $\rightarrow \neg\Diamond_{(C)}\varphi$ (repugnant to esse); "resurrection by divine power" $\rightarrow \Diamond_{(C)}\varphi$ though not $\Diamond_n\varphi$ (not repugnant to esse, though not nomologically possible).

- **Square circle**: Contradictory(φ) $\Rightarrow \neg\Diamond_{(T)}\varphi$ and $\neg D\varphi$.
- **Resurrection**: $\Box_n\neg\varphi \wedge D\varphi \wedge \Diamond_{(T)}\varphi$ (not a contradiction).
- **Accident-without-subject** (as a test case): often $\neg\Diamond_{(T)}\varphi$ if it violates Thomistic metaphysics, though $\Diamond_1\varphi$ may hold.

We will use a classical multimodal formulation to demonstrate 1) contradictory objects are not grounded in being and thus are not possible, and 2) explain why Heisenberg's uncertainty principle does ***not*** entail contradictions and why it does not violate the law of non-contradiction.

1) Contradictory objects: not grounded in being ⇒ not possible

Claim. If a description is contradictory, it lacks the *ratio entis*; therefore it's not possible (and not within omnipotence).

Multimodal Classical Logic (MCL) proof.

1. Assume Contradictory(φ).
2. By Bridge (1) (PNC filter): Contradictory(φ) $\rightarrow \neg\Diamond_{(C)}\varphi$.
3. So $\neg\Diamond_{(C)}\varphi$.
4. By Bridge (4): Contradictory(φ) $\rightarrow \neg D(\varphi)$. **Therefore:** contradictions are **not** $\Diamond_{(C)}$-possible and **not** doable by divine power.

(Example: "square circle" instantiates Contradictory(φ); conclude $\neg\Diamond_{(C)}\varphi \wedge \neg D(\varphi)$.)

We have demonstrated that anything whose description entails a contradiction (like a "square circle") has no footing in reality—it's repugnant to being—so it isn't genuinely possible at all; that's why even omnipotence doesn't "make" contradictions.

2) Why Heisenberg Uncertainty Principle and Quantum Mechanics (HUP/QM) are not contradictory objects

What HUP actually says (nomological):

$$\Box_n \forall S\ (\Delta x(S)\cdot\Delta p(S) \geq \hbar/2)$$

—i.e., given the laws, no state has *both* arbitrarily sharp position **and** arbitrarily sharp momentum. "Sharp" just means **definite**: an observable has a single value with **probability 1** (zero spread). The system is in an **eigenstate** of that observable (variance $\Delta A = 0$). "Not sharp" means it doesn't have a single definite value—outcomes come with probabilities ($\Delta A > 0$).

Key logical point: HUP denies a **conjunction** of sharp values; it does **not** assert any **p** $\wedge$ **¬p** form. So it doesn't trigger the PNC filter.

Multimodal Classical Logic (MCL), consequences:

- From laws to being (Bridge 3): $\Box_n \psi \rightarrow \Diamond_{(C)} \psi$. Hence HUP is at least **metaphysically admissible**.
- HUP is: $\Box_n \neg(\text{Sharp}_x \wedge \text{Sharp}_p)$. This is compatible with each disjunct separately: $\Diamond_n \text{Sharp}_x$ **or** $\Diamond_n \text{Sharp}_p$ (and, idealizing, each entails large dispersion in the conjugate). No contradiction.

Common "looks-like" contradictions, disarmed:

1. **Wave/particle duality.** Properly read as **context-indexed** attributions:

   In context c_wave: $\Box_n \text{ExhibitsInterference}(e)$.

   In context c_part: $\Box_n \text{LocalizedHit}(e)$.

   Not the same respect/time/context $\Rightarrow$ no $\psi \wedge \neg\psi$. (PNC only forbids same-respect collisions.)

2. **Superposition.** $\psi = \alpha|x_1\rangle + \beta|x_2\rangle$ is a **state vector**, not the proposition "$x_1 \wedge \neg x_1$." Before measurement, incompatible outcomes are **in potency**; upon measurement, one becomes **in act**. Classical Metaphysics: potency $\neq$ act, so no contradiction.

3. **"Measurement disturbs = contradiction?"** HUP isn't (just) measurement error; it's a structural constraint from non-commutation ($[x,p]=i\hbar$). But again, it forbids **joint sharpness**, not affirms **p and ¬p**.

There is **no** φ such that physics asserts $\varphi \wedge \neg\varphi$. So Bridge (1) is never triggered against HUP/QM. Instead we have a lawful nomological constraint: $\Box_n \neg(\text{Sharp}_x \wedge \text{Sharp}_p)$, which by Bridge (3) is comfortably $\Diamond_{(C)}$-**compatible**

Heisenberg's uncertainty principle and standard quantum mechanics don't assert contradictions; they impose a lawful limit: you can't have perfectly sharp position and perfectly sharp momentum **together**. That's a prohibition of a joint package, not an affirmation of "p and not-p."

Read from classical modal metaphysics, the laws say "no joint sharpness," which is fully compatible with metaphysical possibility because it doesn't violate the principle of non-contradiction. Hence: contradictions are impossible *simpliciter*, while quantum limits are nomological constraints that sit comfortably within what can be—no threat to being, no contradiction, no collapse.

Whet it's said that "quantum limits are nomological constraints," we mean they're **rules are built into the laws of nature** ($\Box_n$), like the speed-of-light limit or Heisenberg's bound $\Delta x \cdot \Delta p \geq \hbar/2$. Such rules **restrict which packages of properties can be jointly actual** (e.g., "perfectly sharp position AND perfectly sharp momentum"), but they **don't assert a contradiction** like "p AND not-p."

Formally, quantum mechanics says $\Box_n \neg (\mathbf{Sharp_x} \wedge \mathbf{Sharp_p})$—it bans a conjunction. The classical metaphysical modal reads that they do exactly what we want: because laws presuppose being, $\Box_n \psi \rightarrow \Diamond_{(}\mathbf{T}_{)}\psi$ (Bridge 3), the law itself sits **comfortably within what-can-be** ($\Diamond_{(}C_{)}$). There's **no threat to being** (nothing repugnant to *esse*), **no contradiction** (we never get $p \wedge \neg p$ in the same respect/time), and **no modal collapse** (we don't illicitly infer from "ruled out by current physics" to "metaphysically impossible"—that would violate our bridge policy).

Quantum theory does **not** say observation deletes existence. What it says is: **before measurement**, many quantities don't have **definite values**; they're **in potency** (superposed possibilities). **At measurement**, one definite outcome is **actualized**, and conjugate quantities become correspondingly indefinite. In classical metaphysics: the system **exists** the whole time; what changes is which potential **property** becomes **in act**.

That's very different from saying "the particle both exists and doesn't exist," or "it blinks out when you look." Even the dramatic cases (double-slit, "which-path," quantum Zeno) show that **the measurement context alters which properties can be jointly sharp**, not that being toggles on and off. So: quantum limits = lawful constraints on **which determinations can be actual together**—not contradictions, not metaphysical impossibilities, and not a denial of existence.

A **measurement context** mentioned above is the whole setup that fixes **which observable** you're actually measuring—i.e., the basis picked out by the apparatus–system coupling. Call it *CAC_ACA* when the device measures observable AAA (position, spin-z, which-path, etc.).

The **apparatus chooses the questions the world can answer sharply**: a context fixes one maximal set of **mutually commuting** properties to be jointly sharp; incompatible properties cannot be sharp together—so measurement doesn't toggle *being*, it fixes **which potential determinations can be simultaneously in act**.

The reason why context fixes a maximal set is because a "context" is just a commutative subalgebra of observables (a maximal set of mutually commuting operators). Inside that subalgebra, joint sharpness is possible; outside it, it isn't. A measurement device couples to the system with an interaction. That coupling selects **which operator AAA** is being asked about. During and after the interaction, **decoherence** drives the device+system into a state that's effectively diagonal in the **eigenbasis of AAA** (the "pointer basis"). The result is propositions about AAA (and anything commuting with AAA) can become **sharp**; propositions about incompatible BBB do **not**. Rotate the device (Stern–Gerlach along xxx instead of zzz), and you've chosen a **different context**—hence a different family of properties that can be sharp together.

The **Kochen–Specker** style constraint (and Bell/CHSH phenomena) shows you cannot assign sharp, non-contextual values to *all* observables while respecting quantum relations. That's why there is **no single global context** that makes everything sharp: nature forbids it. Extracting perfectly precise information about AAA imprints that info on the pointer. For **incompatible** BBB, the same interaction destroys the phase relations you'd need to read BBB sharply—hence the uncertainty/product bound. It isn't a paradox; it's the price of printing one answer cleanly.

Before measurement, multiple outcomes for AAA are **in potency**. The apparatus **actualizes** one member of the commuting family it selected; incompatible determinations remain unactualized (not "denied existence," just **not in act**). No claim of $p \wedge \neg p$ is ever made, so **no**

contradiction and **no threat to being**: the system exists throughout; only **which determinations are simultaneously in act** depends on context. There is no φ such that physics asserts $\varphi \wedge \neg\varphi$. Bridge (1) is never triggered against HUP/QM. What we have is a lawful nomological constraint, $\Box_n\neg(\text{Sharp}_x \wedge \text{Sharp}_p)$, which by Bridge (3) is comfortably $\Diamond_{(C)}$-compatible. A $\varphi \wedge \neg\varphi$ contradiction would be *repugnantia*: like making a square circle.

The term "repugnant" (*repugnans, repugnantia*) in classical metaphysics is a **technical scholastic term**, not an expression of taste. It became a standard scholastic technical term used broadly in late medieval logic and metaphysics to mean contradictory, incompatible, or contrary to the nature of being or contrary to the nature of things. It was used by Aquinas in metaphysics to describe intrinsic incompatibility, especially when speaking of contradictions in being (e.g. "to be and not to be at the same time is repugnant to reason and reality" (See, ST I, q. 25, a.4 where Gods omnipotence does not extend to things that are *repugnantia* (like making a square circle).

Albert the Great the teacher of Aquinas used *repugnantia* in his commentaries on Aristotle to indicate propositions or natures cannot cohere in the same subject. For example, in his *De Caelo* commentary, he calls mutually exclusive qualities "repugnant in the same subject at the same time." William of Ockham uses it in logic where propositions are impossible if it contains repugnance (contradiction) *Summa Logicae* I.74. Scotus, Suarez, and Cajetan all used the same.

In Aquinas it means **incompatible with being**: a description that, when unpacked, **cannot be instantiated by anything** because it builds in an opposition that destroys the conditions for *esse* (existence/act). Classic formula: what "implies a contradiction" does **not** fall under divine power because it lacks a *ratio factibilis* (the character of something makeable). In practice there are two main kinds:

- **Logical repugnance:** outright **p $\wedge$ ¬p** in the same subject/time/respect ("square circle," "married bachelor").

- **Metaphysical repugnance:** a **structural clash** in the order of being (e.g., "pure potency actually exists," "accident without a subject," "created and uncreated in the same respect"). These reduce to contradiction once the Thomistic notions (act/potency, substance/accident, essence/existence) are applied.

By contrast, **nomological impossibilities** (faster-than-light travel, spontaneous entropy reversal, bodily resurrection by **natural** causes) are **not** "repugnant to being": they're ruled out by our **current laws** ($\Box_n\neg\varphi$), but they don't encode a contradiction and thus aren't blocked by classical metaphysical possibility ($\Diamond_{(T)}\varphi$). In our MCL "repugnant" is the predicate **Repugnant_to_Esse(φ)**, and the key rule is:

Repugnant_to_Esse(φ) $\Rightarrow \neg\Diamond_{(C)}\varphi$ **and** $\neg$**D(φ)** (not metaphysically possible; not within omnipotence).

In Aquinas, to say a description is **"repugnant (to being)"** (*repugnantia ad esse*) means the description **cannot be instantiated by any thing**, because it **builds in an incompatibility** that destroys the *ratio entis*—the very aspect under which something could be. In our notation: **Repugnant_to_Esse(φ)** $\Rightarrow$ there is **no essence** that could realize φ, so $\neg\Diamond_{(C)}\varphi$ and (since omnipotence ranges only over true makeables) $\neg$D(φ).

Here's a clear, three-tier way to see what counts as "repugnant"—and why:

1. **Logical (pure contradiction).**

   Same subject/time/respect, asserting **p** $\wedge$ $\neg$**p** or concept-terms that formally negate each other: "square circle," "married bachelor," "wholly moving and wholly not-moving (same respect/time)." These violate PNC outright.
   $\Rightarrow$ Repugnant_to_Esse(φ).

2. **Metaphysical structural clash (violates what a thing would have to be to be at all).**

These don't look like p ∧ ¬p on the surface, but under classical metaphysics first principles they collapse into it:

o **"Pure potency exists (without act)."** Potency is *of/for* act; to be "pure potency actually existing" is to be both **only-in-relation-to-act** and **already in act**—incompatible.

o **"Created yet uncreated (in the same respect)."** Mutually exclusive per essence.

o **"An essentially-ordered causal series with no first actualizer"** (in the same order of here-and-now dependence): makes the series depend on act that is never present; on classical metaphysical analysis this undermines the actuality the series needs.

In each case, once unpacked, the description **negates the conditions of being** for that subject—hence no *ratio factibilis*. ⇒ Repugnant_to_Esse(φ).

3. **Order-relative (nomological) limits (not repugnant).**

These are **not** contradictions; they're **law-of-nature** exclusions: faster-than-light travel, spontaneous macroscopic entropy reversal, bodily **resurrection by natural causes**. They are $\Box_n\neg\varphi$ (ruled out by the laws acting on their own) but **not** repugnant to being. Hence they can still be $\Diamond_{(C)}\varphi$ if some higher cause (e.g., **D(φ)**) acts. ⇒ Not Repugnant_to_Esse(φ) (so not blocked by $\Diamond_{(C)}$).

Why "repugnant" ⇒ "impossible"? Because **power is ordered to being**. If a "description" fails to present anything that could be—if its very content destroys the conditions for *esse*—then there is **nothing there to be made**. Aquinas's line is classic: what implies contradiction doesn't "fall under omnipotence," not because power is weak, but because **there's no object** (no *ratio factibilis*) for power to actualize.

A quick guide:

- **Same-respect test:** Does φ force **p and not-p in the same respect/time/subject**? If yes ⇒ repugnant.
- **Essence test:** Does φ require an x to **both have and lack** an essential condition of being (e.g., be pure potency yet in act)? If yes ⇒ repugnant.
- **Order test:** Is φ merely **ruled out by current physics** while remaining intelligible as being given a higher cause? If yes ⇒ **not** repugnant ($\Box_n\neg\varphi \land \Diamond_{(C)}\varphi$ consistent).

This also explains why **mystery ≠ contradiction**: e.g., the Trinity, classically parsed (one nature, three Persons), does **not** assert "3=1 in the same respect," so it's **not** repugnant to being, even though it's above what unaided reason can derive.

In the case of quantum mechanics, even with a *perfect* device, **nature itself** won't let position and momentum be *simultaneously exact*. That's because, in quantum theory, "position" and "momentum" are tied to operations that **don't fit together** the way classical quantities do. Technically, their order matters (they **don't commute:** $[x,p]=i\hbar$ [x,p]=i\ hbar), and mathematically the two are connected by a **Fourier transform**: making a wavepacket very **narrow** in space (exact position) automatically makes it very **broad** in momentum, and vice-versa. So the limitation is **built into the structure of states**, not a flaw in the meter.

Any **physical interaction** with an apparatus couples the particle to a device that is tuned to one question (position *or* momentum). That coupling **prepares/selects a context** in which *that* quantity can become **sharp** (definite). Because of the non-commutation, the same interaction **destroys** the phase relations you'd need to also make the incompatible quantity sharp. So the measurement doesn't turn being on/off; it **decides which property can be exact** right now. In **standard QM**, a particle generally **doesn't have** both an exact position and an exact momentum at once; the state simply can't support that. In **Bohmian** or other realist views, the particle may **have** both, but the **theory still forbids** any measurement that makes both jointly exact—the same operational trade-off remains. Either way, there's **no contradiction** (not "p and not-p"); there's a **lawful, structural trade-off**: you can make one exact only by letting the other become fuzzy. In our MCL framework, that's a **nomological constraint** ($\Box_n\neg(\text{Sharp}_x \wedge \text{Sharp}_p)$) that sits comfortably within what-can-be ($\Diamond_\langle T_\rangle$), not a threat to being.

Therefore, in classical metaphysics, a thing is **possible** (possibile) iff its essence is *non-repugnant to being* (*non repugnantia ad esse*). Anything whose description **implies a contradiction** is repugnant to the *ratio entis* (the "notion of being"), so it isn't a *res* at all—it's not even the sort of thing that could be made. Hence Aquinas's famous line: what implies a contradiction "does not fall under omnipotence," because it lacks the ***ratio factibilis*** (the character of something makeable). (ST I, q.25, a.3)

Now how would we formalize chaos? **Let's take tornadoes as contingent natural effects:** when idealized sufficient conditions obtain, $\Box_l$ they occur; across metaphysical space those conditions may or may not obtain (**Cont**). They are produced by **secondary causes** in an order **sustained by the Primary Cause**, and their **chaotic dynamics** limit human predictability while preserving determinism.

Formal Notation:

Language (single-sorted domain with predicates)

- **T(e)**: *e* is a tornado event
- **Nat(e)**: *e* is a natural event
- **Suff_T(e)**: the *idealized* sufficient antecedent conditions for a tornado obtain for *e* (e.g., instability, shear, lift, moisture, vorticity, etc., at the right scales)
- **Causes(X,e)**: X (a set/chain of secondary causes) is a sufficient efficient cause of *e*
- **Sec(X)**: X consists of secondary (created, natural) causes
- **Dep(e,y)**: *e* depends on *y* for its existence/operation (sustaining sense)

- **Prim(G)**: G is the primary cause (ipsum esse subsistens)
- **Cont(e)**: e is contingent (metaphysically non-necessary)

Modalities

- $\Box_m$ / $\Diamond_m$: metaphysical necessity/possibility
- $\Box_l$ / $\Diamond_l$: nomological (law-of-nature) necessity/possibility relative to the actual law set **L**
- (Optional epistemic) K_a φ: agent a knows φ

Intuition: $\Box_l$ quantifies over worlds with the same laws **L**; $\Box_m$ quantifies over all metaphysically possible worlds.

Axioms and principles

(LOI) $\forall x \, (x = x)$
(PNC) $\forall x \, \neg(P(x) \wedge \neg P(x))$
(LEM) $\forall x \, (P(x) \vee \neg P(x))$

(Lawful necessitation for tornadoes)

A1. $\Box_l \, \forall e \, (\text{Suff\_T}(e) \rightarrow T(e))$

If the idealized sufficient atmospheric conditions obtain, then (given the laws) a tornado event follows of nomological necessity.

(Metaphysical contingency of conditions)

A2. $\Diamond_m \, \exists e \, \text{Suff\_T}(e) \wedge \Diamond_m \, \exists e \, \neg\text{Suff\_T}(e)$

Across metaphysical space, the antecedent meteorology may or may not be realized.

(PSR—sustaining sense, for contingent natural events)

A3. $\forall e \, (\text{Nat}(e) \wedge \text{Cont}(e) \rightarrow \exists X \, (\text{Sec}(X) \wedge \text{Causes}(X,e)))$

Every contingent natural event has a sufficient chain/set of secondary causes.

(No self-dependence for contingents)

A4. $\forall e\ (\text{Cont}(e) \rightarrow \neg\text{Dep}(e,e))$

(Primary cause sustains secondary causes and the order)

A5. $\Box_m \exists G\ (\text{Prim}(G) \land \forall e\ (\text{Nat}(e) \rightarrow \text{Dep}(e,G)))$

In classical metaphysics, God (*ipsum esse subsistens*) sustains the existence/operation of the natural order and its effects as primary cause, without replacing secondary causes.

(Act–potency schema for natural effects)

A6. $\forall e\ (\text{Nat}(e) \rightarrow (\text{Pot\_T}(e) \leftrightarrow \Diamond_l\, T(e)))$

Natural "potency" to T is just nomological possibility for T under **L**.

(Chaos—sensitive dependence, idealized)

Let **D** be the atmospheric dynamical system under **L**.

A7. (Chaos(D)): For any initial state neighborhood U and any macro-outcome predicate O, there exist s,s'∈U such that O holds for the trajectory from s and ¬O holds for the trajectory from s'.

Encodes sensitive dependence/topological mixing at the forecasting scale of interest.

(Finite epistemic precision)

A8. $\forall a\ \exists \varepsilon > 0$ such that *a*'s pre-event measurements localize initial state only to an ε-neighborhood.

Theorems / derivations

T1. Lawful-but-contingent tornadoes.

From **A1** and **A2**:

1. From **A2**, $\Diamond_m \exists e\ \text{Suff\_T}(e)$. In some metaphysically possible world (with **L** fixed), the sufficient conditions obtain.
2. By **A1**, in any **L**-world where Suff_T(e), we have T(e). So $\Diamond_m \Diamond_l \exists e\ T(e)$.
3. From the second conjunct of **A2**, $\Diamond_m \exists e\ \neg\text{Suff\_T}(e)$. So in some metaphysically possible world the sufficient conditions fail, hence ¬T(e) there (given **A1**'s direction).

4. Therefore $\Diamond_m \exists e\, T(e) \wedge \Diamond_m \exists e\, \neg T(e)$ and so tornadoes are **not** $\Box_m$-necessary: **Cont($\exists e\, T(e)$)**.

Conclusion: Tornadoes are **nomologically lawful** ($\Box_l$-necessitated) by sufficient conditions) yet **metaphysically contingent** (not $\Box_m$).

T2. Secondary-cause sufficiency (Classical Metaphysics PSR).[31]

Given a particular tornado event τ:

- From **T1**, Cont(τ).
- From **A3**, $\exists X\, (Sec(X) \wedge Causes(X, \tau))$.

Conclusion: The tornado is explained by an ordered set/chain of **secondary causes** (e.g., thermodynamic gradients, shear, mesoscale boundaries), not by chance in the sense of lawlessness.

T3. Primary/secondary concurrence (sustaining causality).

- From **A5** and Nat(τ), Dep(τ,G).
- Together with **A3**, τ both **depends on G** (primary, sustaining cause) **and** proceeds **through X** (secondary, natural causes).

Conclusion: No competition: the same effect is caused *per se* by secondary causes within the created order and *per modum causae primae* by the primary cause sustaining the order.

T4. Deterministic yet practically unpredictable (chaos).

- From **A7 (Chaos)** and **A8 (finite precision)**, before the event time, no finite agent a can rule out macroscopically distinct outcomes that hinge on sub-ε perturbations.

---

31 The Principle of Sufficient Reason (PSR) in classical metaphysics is grounded in ontology. It is not a rationalist notation such as that of Leibziz's PSR. For Leibniz, PSR is a rationalist axiom or rule of thought or formal logic, for every fact there must be a reason why it is so. For classical metaphysics, it is a law of being like the law of non-contradiction. For classical metaphysics, for everything that comes to be or that is contingent, there must be a sufficient reason (cause) in virtue of which it exists; this reason is either in another (an efficient cause) or ultimately in God, whose essence is existence itself. It is not simply logical, but metaphysically grounded. The reason for any contingent being lies either in another being (its cause) or ultimately in a necessary being *(ipsum esse subsistens)*

Formally (schematic):

- $\forall a$ (FinitePrecision(a) $\land$ Chaos(D) $\rightarrow \neg\Box(K_a$ pre-t (Suff_T(e) obtains exactly)))
- Yet by **A1** the mapping from exact Suff_T to T is $\Box_l$-deterministic.

Conclusion: Tornadoes are **deterministic under the laws** but **epistemically chaotic** to finite agents.

Notes on Classical Metaphysics:

- **Aristotle ("for the most part"):** A1–A2 capture that natural effects follow ordered principles but depend on contingent convergences—not strict periodic necessity.
- **Aquinas (ST I, qq. 103–116):** A3, A5 express **secondary causes** genuinely producing effects within an order **conserved by the First Cause**. Tornadoes are not "by chance" in the sense of lawlessness; they arise when multiple ordered causes intersect contingently.
- **Act & potency:** A6 formalizes that the atmosphere has a *potency* for tornadogenesis ($\Diamond_l$ T) that is *actualized* when sufficient conditions obtain.
- **Chaos $\neq$ denial of causality:** A7–A8 show why forecasting is limited without impugning lawfulness.

Therefore tornadoes are contingent natural effects: when idealized sufficient conditions obtain, $\Box_l$ they occur; across metaphysical space those conditions may or may not obtain (**Cont**). They are produced by **secondary causes** in an order **sustained by the Primary Cause**, and their **chaotic dynamics** limit human predictability while preserving necessity and Providence.

Now let's argue from chaos to *Ipsum Esse* or the Primary Cause using a two-column modal argument.

Symbol/Predicate Notation

- $\Box_m/\Diamond_m$: metaphysical necessity/possibility; $\Box_l/\Diamond_l$: nomological (laws **L**) necessity/possibility
- **Nat(e)**: e is a natural event; **T(e)**: e is a tornado event
- **Exact(D,t,s)**: in dynamical system **D** (atmosphere under **L**), at time t the exact state is s
- **Det$_m$(e)**: e is determinate in respect m; here m^{**} = "exact state along the chaotic trajectory τ"
- *Cause*$(X,e)^{**}$: X is a *per se* (concurrent/sustaining) cause of e
- **Sec(X)**: X is a (possibly plural) secondary cause within nature
- **Can$_m$(X)**: X can ground proportionately the determination in respect m
- **Dep(x,y)**: x depends (sustainingly) on y
- **Ess(x)=Esse(x)**: x's essence is its existence
- **Ex(x)**: x exists; **Eternal(x)**: x is eternal (no temporal succession)

Axioms & Principles Used (from prior setup)

- **A1 (Lawful necessitation)**: $\Box_l \forall e$ (Suff_T(e) $\rightarrow$ T(e)).
- **A7 (Chaos)**: Sensitive dependence/topological mixing for atmospheric **D** at forecast scales.
- **LOI/LEM**: Identity/Excluded Middle (determinacy of actuality).
- **C1 (Deterministic chaos exactness)**: $\Box_l \forall t \exists !s$ Exact(D,t,s).
- **C2 (Proportionate causality/adequation)**: $\forall e \forall m (\text{Det}_m(e) \rightarrow \exists Y(\text{Cause}^*(Y,e) \land \text{Can}_m(Y)))$.
- **L3 (Inadequacy of purely secondary causes for m*)**:** $\forall X(\text{Sec}(X) \rightarrow \neg\text{Can}_m^*(X))$.
- **NR (No *per se* regress)**: Every *per se* sustaining series is wellfounded (terminates).
- **EE (Essence–Existence bridge)**: *Per se* terminus proportionate without limit $\Rightarrow$ Ess=Esse, $\neg\exists y$ Dep(x,y), $\Box_m$Ex(x).

Two Column Argument (Left: Formal; Right: Explanation)

Modal Logic Steps	Explanation
1) **A7**. Chaos(D).	The atmosphere under laws **L** exhibits sensitive dependence; tiny state changes can flip macrooutcomes.
2) **C1**. $\Box_l \, \forall t \, \exists! s \, Exact(D,t,s)$.	In a classical deterministic model, at every time there is one exact state; chaos is not lawless.
3) From 1–2 + **LOI/LEM**: $\forall t \, Det_m{}^*(\tau(t))$.	The actual world instantiates a single, fully determinate trajectory τ; it's "this exact path," not an indeterminate blur.
4) **C2** with m*: $\forall t \, \exists Y \, (Cause^*(Y,\tau(t)) \wedge Can_m{}^*(Y))$.	Adequation: if an effect is determinate in a respect, there must be a concurrent cause proportionate to that determination.
5) **L3**: $\forall X \, (Sec(X) \to \neg Can_m{}^*(X))$.	No sum/chain of potencylimited secondary causes suffices to ground the unbounded exactness that chaos demands.
6) From 4–5: $\forall t \, \exists Y \, (Cause^*(Y,\tau(t)) \wedge \neg Sec(Y) \wedge Can_m{}^*(Y))$.	Therefore the needed concurrent cause at each instant is beyond the order of merely secondary causes.
7) **NR** applied to the *per se* chain at t: $\exists G \, (Cause^*(G,\tau(t)) \wedge Per\,se\text{Terminus}(G) \wedge Can_m{}^*(G))$.	*Per se* sustaining dependence must terminate in a cause that does not itself require concurrent support, and that is proportionate to m*.
8) **EE** on G: $Ess(G)=Esse(G) \wedge \neg \exists y \, Dep(G,y) \wedge \Box_m Ex(G)$.	Bridge: a *per se* terminus proportionate without limit (no potency limiting its act) is *ipsum esse subsistens* and exists necessarily (not by another).
9) From no potency in G $\Rightarrow$ no change $\Rightarrow$ **Eternal(G)**.	Standard Thomistic lemma: pure act $\Rightarrow$ immutability $\Rightarrow$ eternity (no before/after in G).
10) Conclusion: Deus est *ipsum esse subsistens per se aeternum.*	Naming G as Deus: necessary, selfsubsisting existence itself, eternal.

Why this does not beg the question

1. We do **not** assume a Primary Cause exists (A5 is unused).
2. The scientific core—deterministic evolution + sensitive dependence—grounds steps (1–3).
3. The only contested lifts are **C2 (adequation)**, **NR (wellfounded *per se* series)**, and **EE (terminus ⇒ Ess=Esse)**—all standard classical metaphysical theses defensible independently of theology.
4. The pressure point is **L3**: chaos requires **unbounded** exactness that no potencylimited secondary causes can proportionately ground *per se*. Deny L3, and you owe an account of how bounded causes ground an unbounded determination without collapsing into an ungrounded *per se* chain, contra **NR**.

On L3 we can replace "unbounded exactness" with "perinstant actualization among the indefinitely many equipollent trajectories" in a discrete physics variant; the proportion/termination argument still goes through.

Modal scope: $\square_l$ ranges over **L**worlds; $\square_m$ over metaphysical space. Necessity of **G** is $\square_m$, not merely $\square_l$.

Metaphysical upshot: **Chaos sharpens**—not weakens—the need for a proportionate, nonderivative **concurrent ground of determinate actuality**.

Deny L3, and you owe an account of how bounded causes ground an unbounded determination without collapsing into an ungrounded *per se* chain, contra NR.

What L3 says:

L3: Purely **secondary causes** (each finite, potency-limited) are **not proportionate** to the "unbounded exactness" needed to fix a single chaotic trajectory (m^*).

Intuition: chaotic determinacy requires distinctions as fine as you like; no *finite* package of causes contains that degree of determination.

If you deny L3, you owe one of the following stories (and each runs into trouble):

1. **Finite-package story (underdetermination).**

    Claim: Some *finite* set of secondary causes **X** already grounds the exact trajectory.

    Problem: By proportionate causality (C2), the cause must contain the effect's determination "in some way." But finite, potency-limited causes can only fix **finitely** many distinctions. That leaves multiple admissible trajectories compatible with X—so the actual path is **underdetermined** (contradicting Step 3: the world in fact instantiates one exact path). To escape, you'd have to insert a brute selector ("it just happens"), which violates C2/PSR.

2. **More-and-more story (per-se infinite regress).**

    Claim: Add further secondary causes, ever finer, to reach the needed exactness.

    Problem: If those causes are required **concurrently** to sustain the determination "here-and-now," you've built a **per-se** chain that has no terminus. That's precisely an **ungrounded per-se chain**, which **NR** forbids. (If you instead say the extra causes come **earlier in time**, that's a *per accidens* series and can't ground a present determination as such.)

3. **Law-only story (nomic generality isn't individuation).**

    Claim: The **laws** themselves pick out the exact state.

    Problem: The same law set **L** is compatible with many nearby states producing different macro-outcomes (that's chaos). Laws are **general**; they don't individuate *which* exact initial value obtains. You still need a proportionate **actualizer** for the particular.

4. **Randomness story (gives up the setup).**
   Claim: Indeterministic "chance" selects the state.
   Problem: That abandons **C1** (deterministic chaos) and changes the target. If you keep C1, probabilistic talk is epistemic only and doesn't supply an ontic selector.

Why this collapses into "ungrounded *per-se* chain"

- If **finite** secondary causes don't suffice (1), and **only more** of the same kind could (2), then the determination m* would depend on an **actually infinite** hierarchy of concurrent potency-limited causes. That hierarchy has **no self-sufficient member** to confer actuality; it's dependence "all the way up."
- A *per-se* series without a **terminus** is exactly an **ungrounded series**—**contra NR** (the principle that per-se sustaining series are well-founded).

The core trilemma (a succinct version):

Either the exact chaotic trajectory is grounded by:

- **(A)** a **finite** secondary cause set → **underdetermination** or brute selection (violates C2/PSR), or
- **(B)** an **infinite** per-se stack of secondary causes → **ungrounded** (violates NR), or
- **(C)** a **terminating cause proportionate without limit** → not potency-bounded, i.e., **pure act** → **ipsum esse subsistens**.

To **deny L3** coherently, one must give a precise model where **bounded** causes ground **unbounded** determination **without** (B) and **without** smuggling in brute selection (A). That's the debt the line points to.

In summary, although the atmosphere in the case of a Tornado are under laws **L** and exhibit sensitive dependence; tiny state changes can flip macrooutcomes. In a classical model, at every time, there is one exact state; therefore, chaos is not lawless. The actual world instantiates a single, fully determinate trajectory τ; it's "this exact path," not an indeterminate blur. If an effect is determinate in some respect, there must be a concurrent cause proportionate to that determination. No sum/chain of potencylimited secondary causes suffices to ground the unbounded exactness that chaos demands. Therefore, the needed concurrent cause at each instant is beyond the order of merely secondary causes. *Per se* sustaining dependence must terminate in a cause that does not itself require concurrent support, and that is proportionate to m*. This requires a *per se* terminus that is proportionate and without limit (no potency limiting its act).

Consequently, this is *ipsum esse subsistens* and exists necessarily (not by another). pure act $\Rightarrow$ immutability $\Rightarrow$ eternity (no before/after in God). God is *ipsum esse subsistens aeternum et necessarium* (the very act of being selfsubsisting, eternal and necessary).

CONCLUSION

Now let us return to our Socratic dialogue in conclusion. Let's pull together our thoughts on the nature of Being and Necessity, the role of experience for Hume, Kant's transcendental philosophy, the rigid designators and possible worlds of Kripke, and a reformulation of the Synthetic *a priori* based in classical metaphysics and noetics.

In introducing the dialogue, let's quickly review a few points. Aristotle gave a careful treatment of *chance* (τύχη, *tychē*) and *the fortuitous* (ἀπὸ τοῦ αὐτομάτου, *apo tou automátou*) in **Physics II**, distinguishing them from both necessity and purposive causality. Aristotle argued that most events in the natural world happen **either by necessity (cause and effect)** or **for the sake of an end (teleology)**. But sometimes events occur that do not fit neatly into these categories. For example, you might go to the marketplace to buy food and unexpectedly meet a debtor who pays you back. The repayment wasn't *for the sake* of your trip, but it still happened. Such occurrences he called **by chance** or **fortuitous**.

For Aristotle, the *fortuitous* (**Automaton**) is a broad category: events that happen incidentally (παρὰ συμβεβηκός, *para symbebēkos*) rather than essentially. Example: A stone falls downward by its nature (*physis*), but if it kills someone, that death is not its natural end; it is fortuitous.

Chance (*Tychē*) is a narrower notion, reserved for fortuitous outcomes in the sphere of **human action and deliberation**. Only beings capable of choosing ends and acting for them (humans, or in a looser sense, rational agents) can properly have "luck" (*tychē*). So, finding treasure while digging a hole is "by chance," but only because the digger had a goal in mind. Nature by itself does not literally have "luck." Aristotle did not reduce chance to mere illusion or ignorance. He considered it a **real kind of cause**—but not a primary one. It is a *per accidens cause* (incidental cause): it arises from the intersection of independent causal chains. For instance: a man goes to market to sell produce, another goes to market to meet a friend; by coincidence they meet. Each had a reason, but their meeting was incidental.

This account allows Aristotle to preserve **teleology** (most things happen for a reason) while admitting genuine contingency. He rejected the atomists' view (e.g. Democritus) that "chance" is simply ignorance of necessity. For Aristotle, chance is *rare* and *irregular*, but still a natural feature of the world. Indeed, genetic drift, incidental mutations, and biological evolution would fall under Aristotle's view of *fortuitousness*.

Aristotle viewed chance and luck not as fundamental causes, but as **incidental outcomes of intentional or natural processes**. They are real, but secondary—arising from the intersection of independent causal chains, particularly in human life where goals and purposes are involved. Aristotle defines chance as a *per accidens* **cause**: the incidental meeting of independent causal chains. These chains are real, chance events are real. But they are **rare, irregular, and not for the sake of something** — unlike events governed by natural necessity or human purpose.

Aristotle never says the Prime Mover directly controls chance. Rather, since the Prime Mover is the **ultimate final cause of the universe's order**, all things are indirectly referred back to it, including chance events. But this is not determinism in the modern sense because the Prime Mover does not *efficiently* determine every detail. Chance remains a feature of the world precisely because multiple purposive actions or natural tendencies can intersect in unintended ways.

For **atomists like Democritus**, "chance" was just ignorance of necessity; everything is strictly determined. For **Aristotle**, chance is not reducible to necessity: it is a **genuine feature of reality**, though subordinate to the general order. The Prime Mover ensures the cosmos has order, intelligibility, and teleology. But within that framework, contingency and chance still occur. Thus for Aristotle, his system is **not deterministic** in the strict sense. It preserves a structured universe without eliminating contingency.

Aristotle did not argue that chance events are "controlled" by the Prime Mover in the way determinism suggests. Instead, the Prime Mover is the ultimate cause of cosmic order, while chance arises as incidental outcomes within that order. The Prime Mover guarantees intelligibility and final causality, but **leaves space for genuine contingency** in the sublunary world. It leaves room for biological changes, whether those changes are by natural selection or purely accidental convergence or genetic drift.

If everything down to the trailer park strike in the previous Tornado example were "controlled" by the Prime Mover, that would collapse into a kind of determinism. Aristotle avoids this by insisting that **chance is real** and *per accidens* **causality is not reducible to necessity**. The Prime Mover secures intelligibility at the level of the whole, but not predestination of particulars. For Aristotle, the tornado itself follows natural necessity, but the fact that it hits *this* trailer park rather than another is a **chance event** — not directed by the Prime Mover either directly or indirectly. The Prime Mover provides the overall framework of cosmic order, but allows for genuine contingency in the sublunary world.

Classical metaphysics reformulated Aristotle's view, God (*the Ipsum Esse Subsistens*) is both **first cause** and **final cause**. Nothing at all falls outside divine causality. Classical metaphysics keep Aristotle's idea that chance is real at the level of *secondary causes*. From our human perspective, the tornado hitting *this* park instead of that one is truly a chance outcome. Yet, for the Parisian and Scholastic philosophers of the 13-17th centuries, they insisted that even these chance events occur within the scope of God's providence. They are not "outside" God's knowledge or governance.

Aquinas argues in *Summa Theologiae* I, q.22, a.2: "Nothing happens outside the order of divine providence." But God permits contingency by **giving creatures genuine causal powers**, so that their independent actions can intersect. God does not *will* every particular outcome *as such* (e.g., He does not directly will "this tornado destroys this trailer park"), but He permits it within a providential order that allows for real secondary causation. For **Aristotle there is NO determinism**. The Prime Mover sustains order but does not predetermine particular outcomes. Chance is ontologically real. **Aquinas has a** stronger determinacy at the universal level (everything is within God's providence). But not *fatalism*: Aquinas insists God's providence *includes* genuine contingency. He says God's plan includes not only necessary causes but contingent ones, and it is "more perfect" that the world contains both.

For the Atomists, the atoms and natural laws are deterministic and there is no chance or contingency. The atomist framework would leave little room for evolution, or small incidental changes over time. For Aquinas, what is chance to us is foreknown to God, but still genuinely contingent. Meaning, evolutionary changes, although contingent in themselves are foreknown to God and still Providential and under his control for Aquinas.

Contemporary Thomists following classical natural philosophy, like Michael Dodds in works like *Unlocking Divine Action* (2012), stresses that **Aquinas's understanding of chance** is not *opposed* to providence. Rather, it arises from the **interaction of real secondary causes**. Dodds would say a tornado hitting one trailer park as opposed to another is not meaningless randomness; it reflects the **real autonomy of natural causes**. God allows these causes to operate according to their own natures, and their intersections produce chance events. Thus, tornado damage is not "micromanaged" by God but still occurs under providence.

Dodds applies the same reasoning to **biology and evolution** where genetic mutations and variations are **chance events relative to the organism's purpose**. A chance mutation might not be "for the sake of" survival in itself. Yet they are not "uncaused" — they result from physical processes (radiation, copying errors, etc.). Chance here allows for the **open-endedness of creation**. New forms and species arise not as pre-programmed determinism but through contingent processes that God permits within providence. Such events are not directed as such by God, but they exist within God's providential order, which intentionally includes contingency as a feature of creation's perfection. In his language: chance in evolution "does not lie outside God's plan, but within it, as part of the perfection of providence."

Setting: A circle of philosophers at a long table, with storm clouds outside and news of a tornado strike on the horizon. A storm rumbles in the distance. The philosophers gather to debate.

Atomist: "Look there—storm and tornado! Both are nothing but atoms swirling in the void. $\Box(A \to T)$: if antecedent atomic conditions hold, then the storm or tornado follows with necessity. There is no miracle, no providence. The gods, if they exist, do not intervene."

Socrates: "So you claim everything is determined?"

Atomist: "Exactly so, Democritus says. Epicurus allowed for the *clinamen*, an atomic swerve, $\Diamond(\neg p)$, but that is only brute randomness. Chance is not real cause, only ignorance."

Aristotle: "You reduce everything to necessity or blind chance. But chance is real in another sense: *per accidens*. $\Box(A \to \text{Tornado})$, yes. But $\Diamond(\text{Strike A}) \land \Diamond(\text{Strike B})$. Which park is hit depends on independent causal lines intersecting. That's contingency—not ignorance, not randomness, but a real feature of nature."

Hume: "Aristotle, you smuggle in necessity. All you ever see are sequences—never the 'must.' To say $\Box(A \to \text{Storm})$ is just custom: we're conditioned to expect B after A. No real necessity. And you, Atomist, are worse—you pretend you can see necessity in atoms, but you can't."

Kant (snapping): "David, your skepticism destroys science. If necessity is just habit, you can't justify a single law. $\Box(A \to B)$ is not custom—it's a category of understanding, a synthetic *a priori*. Without it, no experience of nature would be possible. But we must admit: miracles, if they occur, are noumenal, outside the domain of possible knowledge."

Aristotle: "Hume you're mistaken, my friend. Nature acts not only by necessity but also for ends. The storm arises necessarily, $\Box(A \rightarrow$ Storm$)$. Yet the tornado striking this park rather than that is *per accidens*, contingent: $\Diamond P1 \wedge \Diamond P2$. Chance is real where causal lines intersect, not ignorance or blind randomness. But as for miracles, no: nature's powers are not suspended."

Socrates: "So you preserve contingency but not divine intervention?"

Aristotle: "Yes. The Prime Mover governs the cosmos as final cause, but does not direct every storm or tornado strike."

Plantinga: "Aristotle, you speak of chance, but I say contingency is better explained by possible worlds. Imagine W1 where Jesus calms the storm, W2 where He does not. Both are logically possible. God, in His providence, actualizes one. Once chosen, $\Box(\text{W\_actual} \rightarrow \text{all within W})$."

Aristotle: "But then contingency vanishes once the world is chosen! That collapses chance! Once God chooses, no real contingency remains."

Plantinga: "Wrong. Contingency is preserved at the modal level: $\Diamond p$ even if, in W*, $\neg p$. At the modal level, contingency remains. $\Diamond(\text{Storm calmed}) \wedge \Diamond(\text{Storm continues})$. Providence is preserved without contradiction."

Socrates: "Yet is this not determinism once God selects the world?"

Plantinga: "Determinism at the level of actualization, yes, but freedom at the level of world-choice."

Kripke: "Let me cut in. Without rigid designation, this debate collapses. The name 'Jesus' rigidly designates the same individual across all possible worlds. In some, $\Box(\text{Jesus is human})$, but $\Diamond(\text{Jesus is not a carpenter})$. He is essentially human ($\Box$), accidentally carpenter ($\Diamond$). My semantics gives you clarity on essence and modality. So too with the storm: W1 has calming, W2 does not. I give you the semantics; others may debate metaphysics."

Socrates: "So you give us clarity of reference, but not an account of causation or providence?"

Kripke: "Just so. My task is to show that essence is real in modal logic. The rest I leave to metaphysicians."

Heisenberg: "All of you metaphysicians forget physics. Tornado paths aren't determined like clockwork. Quantum fluctuations + chaos make $\Diamond$(Strike A) and $\Diamond$(Strike B) real, objective possibilities. Determinism is dead. Potentiality is built into nature itself—Aristotle's *dynamis* reborn in quantum mechanics. At quantum scales, $\Box$ determinism fails. The tornado's path—Park A or Park B—is not fixed but contingent on quantum fluctuations. The world is structured potentiality: $\Diamond P1 \wedge \Diamond P2$ until measured or realized. Aristotle's *dynamis* lives again! Yet as for Jesus calming the storm, physics offers no account. That is beyond science."

Aristotle (smiling): "So you too see potency until act?"

Heisenberg: "Indeed, though I call it probability amplitudes."

Atomist: "So you've just replaced determinism with randomness."

Heisenberg: "Not randomness—structured indeterminacy, governed by probabilistic law. That's stronger than your void-and-swerve nonsense."

Aquinas: "Hear me, friends. You all speak partially true. The sea has real potency for calm and storm. By secondary causes, $\Box(A \rightarrow \text{Storm})$. But God as *ipsum esse subsistens* sustains being itself. In the miracle, God directly actualizes the sea's potency for calm—no law is broken, only bypassed. The tornado too arises by necessity, yet where it strikes is contingent *per accidens*. Chance is real, but all is within providence."

Dodds: "Physics gives you indeterminacy, but no providence. Thomism integrates both. Natural laws—statistical or deterministic—are real. Chance is real in secondary causes. But God sustains both. Miracles are possible as rare direct acts. Tornado paths, genetic drift, quantum jumps—they're contingent at the level of finite causes, but still within providence. That's the only way to hold law, chance, and divine action together."

Plantinga: "Thomas, you sound convincing, but your ontology is heavy. I'll put it cleaner: there are many possible worlds—W_1 where the storm continues, W_2 where Jesus calms it. Both are possible. God actualizes one. Once W* is chosen, everything is fixed. Contingency remains because another world could have been actual. That's providence without contradiction."

Aristotle: "But that collapses chance! Once God chooses, no real contingency remains."

Plantinga: "Wrong. Contingency is preserved at the modal level: $\Diamond p$ even if, in W*, $\neg p$."

Klima: "Plantinga my friend, in modal terms: $\Box(A \rightarrow \text{Storm})$, $\Diamond(\text{Calm via miracle})$. Possibility is grounded not in abstract descriptions but in real potencies of beings. Contingency = potency not yet actualized."

Dodds: "And let me update this in modern light. Tornado strikes, genetic drift, quantum events—these are chance relative to secondary causes. God sustains the laws of nature, but also wills contingency as part of creation's perfection. Chance is not outside providence; it reveals its richness. In divine governance, $\Box(\text{Providence includes chance})$."

Heisenberg (nodding): "So my quantum indeterminacy finds a metaphysical home in your Thomism."

Socrates: "Let us test each view. Atomist, you deny meaning, leaving us with determinism or brute randomness. Aristotle, you preserve chance but deny miracle. Plantinga, you give us divine choice but collapse contingency once the world is chosen and reduce necessity to abstract modal descriptions. Kripke, you give semantics but no ontology. Heisenberg, you give us indeterminacy, but not providence. Aquinas and Klima, you give being itself as sustaining, chance in secondary causes, and miracles as direct actualization. And Dodds, you show this works even with quantum science."

Aquinas: "Thus the storm calmed is a miracle: not a contradiction, but God directly actualizing a potency. The tornado strike is chance *per accidens*, contingent within natural laws. Both events are providentially sustained, without collapsing into determinism."

Dodds: "Exactly. Divine action is possible without undermining natural law. Chance and contingency remain real, and providence includes them, not by eliminating them but by sustaining them in being."

Kripke (smiling): "And my semantics shows how rigid designation and essence keep the identity of beings stable across worlds. But Aquinas grounds those essences in being itself. Together, we secure clarity and metaphysics."

Socrates (concluding): "So the final harmony is this: against the Atomists, chance is real, not ignorance or brute randomness. Against Aristotle, divine action is possible without denying natural law. Against Plantinga, contingency is preserved not only modally but ontologically. Against Kripke, semantics is deepened by metaphysics. Against Heisenberg, quantum indeterminacy is given a providential frame. Thus, in Aquinas, clarified by Klima, supported by Dodds, and sharpened by Kripke's semantics, we find the most complete account: divine providence sustains natural laws as *ipsum esse subsistens*, miracles actualize potencies directly, and chance and contingency flourish within creation's order."

Hume: "Wait Socrates! Gentlemen, you speak of necessity and cause as though they were visible to the eye. But all we ever observe is one event following another. Fire warms, the storm follows pressure changes, but we never perceive a *must*. All that $\Box(A \to B)$ means is that we are accustomed to seeing A conjoined with B. Causality is not discovered in nature, but in our habits of expectation. A Tornado strikes here or there; it is only expectation, not metaphysical truth."

Socrates: "So, David, you dissolve causality itself into custom?"

Hume: "Indeed. $\Box(A \to B)$ is nothing but repeated experience of A and B conjoined, giving rise to belief. There is no rational ground to infer necessity."

Kant: "Hume, your skepticism again awakens me from my dogmatic slumber! You are right: necessity is not derived from experience. If causality were nothing but habit, science would collapse into mere custom. You are right that necessity is not derived from mere sense impressions — we never see necessity with the eyes. But it is a condition of the possibility of experience itself. Causality is a category of the understanding: a **synthetic *a priori* judgment**. We must presuppose $\Box(A \to B)$ for the phenomenal world to be intelligible. Storms and tornadoes are governed by natural laws not because we see necessity, but because the mind imposes lawfulness on appearances."

Socrates: "So necessity is not in things, but in our framework for knowing them?"

Kant: "Exactly. Things-in-themselves remain unknowable. Contingency and necessity apply only to phenomena under our categories."

Aquinas: "Here, both of you glimpse the truth, but each stops short. David, you are right that our awareness of causality arises through repeated encounters with events. We do not intuit necessity in a single instance; we recognize it as we see convergence across many. Yet it is not mere habit or belief. Through abstraction, the intellect discerns universality and necessity from the very *natures* of things. When the storm follows from pressure and wind, the mind sees more than succession: it sees that effects cannot exist without causes, because of what beings are.

And Immanuel, you are right too: the intellect has an active role. It does not simply register impressions, but forms universal concepts and necessary judgments. Yet the mistake of your transcendental philosophy is to relocate necessity into the subject, as if the mind imposed lawfulness upon appearances. Necessity is not a mental projection. It flows from being itself, sustained by *ipsum esse subsistens*. Universals and necessary propositions are true because they mirror the order of reality, not merely the structure of our faculties. You rightly see necessity is universal and not derived from experience. But you relocate it in the mind instead of in being. $\Box(A \rightarrow B)$ is not imposed by us but flows from the natures of things."

Klima (nodding): "And in modal terms: Hume says $\Diamond(A \land \neg B)$ always remains open, because necessity is only habit. Kant says $\Box(A \rightarrow B)$ holds because the mind structures phenomena. We say $\Box(A \rightarrow B)$ holds because being itself grounds necessity in essence and potency. Possibility and necessity are grounded ontologically in being, not epistemologically in our psychology or categories."

Kant: "But Thomas, how can you claim knowledge of things-in-themselves? My critical philosophy restricts knowledge to phenomena."

Aquinas (responding, firmly): "Immanuel, you mistake the order of things. Yes, the intellect contributes, but not by *imposing* necessity. It contributes by **abstracting universals** and discerning necessity from the way things are. We begin with the senses: *nihil est in intellectu quod non prius fuerit in sensu* — there is nothing in the intellect that was not first in the senses. The mind is a *tabula rasa*, written upon by the nature of things.

From repeated encounters, the intellect abstracts what is universal and necessary. That is precisely what your 'synthetic *a priori*' aims at — but you place it in the subject. In truth, it arises from the convergence of experience with the ontological structure of beings. The universal and necessary judgments we form are not projections, but insights into *esse* — the act of being that grounds natures and their causal relations."

Kant: "Thomas, my critical philosophy and the Transcendental Deduction prove that the categories determine what is necessary and universal."[32]

Aquinas (pressing further): "You see, transcendental philosophy is misguided not because such categories don't exist, but because you **misplace them**. You put them in the subject, as if the mind manufactures necessity. In reality, the categories are grounded in being itself. The mind does not legislate for nature; it recognizes the lawfulness inscribed in being by its Creator."[33]

32 Kant, in contrast, posits that categories are *a priori* forms that structure phenomena:

1. **Quantity:** unity, plurality, totality.
2. **Quality:** reality, negation, limitation.
3. **Relation:** inherence/subsistence (substance/accident), causality/dependence, community.
4. **Modality:** possibility/impossibility, existence/non-existence, necessity/contingency.

For Kant, these are not derived from experience; they are imposed *by the mind* on appearances. They do not describe things-in-themselves (noumena) but only how phenomena must appear to us.

33 Aristotle listed ten categories *(praedicamenta)* as the most general modes of being:

1. **Substance** *(ousia)* – what a thing is.
2. **Quantity** – how much, discrete or continuous.
3. **Quality** – what kind, e.g., hot, white.
4. **Relation** – relative to another, e.g., larger, father.
5. **Place** – where.
6. **Time** – when.
7. **Position** – posture or arrangement.
8. **State** – having, condition (armed, clothed).
9. **Action** – doing, e.g., cutting, burning.
10. **Passion** – being acted upon, e.g., being cut, being burned.

For Aquinas, these categories are not mental constructs but **ontological modes of being**. Universals and

Aquinas: "We know being *qua being* by abstraction, not by intuition of noumena. The act of existence (*esse*) is evident in every being. From this, we see that universal and necessary propositions are not fictions of the mind but rooted in the order of being. When I say □(triangle → three sides), this is not a construct but a reflection of essence. Likewise, chance in tornado strikes is real because finite causes act contingently. Providence sustains both."

Hume: "Yet your 'esse' is unseen! We perceive impressions only. The claim that God sustains being is mere speculation."

Klima: "No, David. Your radical empiricism cannot explain why experience has intelligibility at all. If causality is mere habit, you cannot justify science. Aristotle, Aquinas, and even Heisenberg agree that potency and act, necessity and chance, are structures of reality, not mere psychology. Without being, your skepticism consumes itself."

Dodds: "Immanuel, by placing necessity in the subject rather than in being, you sever science from metaphysics. But modern physics — quantum indeterminacy, chaos theory — shows contingency is real in nature itself, not just in our categories. Classical philosophy, rooted in being, provides a better framework: chance and necessity are ontological, yet encompassed in providence."

Heisenberg (nodding): "Indeed, I found Aristotle's potency a better analogy than Kant's categories for quantum probability. Potentiality is not in the mind but in matter itself."

Aquinas: "So, Immanuel, your categories are not false, but misplaced. You put them in the understanding, as if the mind manufactures necessity and causality. But necessity and causality belong to being. The intellect abstracts them from things through experience and recognizes them as universal and necessary because they reflect what is real. By making the mind the lawgiver of nature, you sever science from reality. By grounding universality in being, we preserve both experience and metaphysics.

Immanuel, you grasp the problem well, but you misplace the solution. The intellect indeed contributes, but not by legislating necessity.

necessities abstracted from experience reflect these real modes.

It contributes by **abstracting universality and necessity from being as encountered in experience**. □(A → B) is not a form projected by us, but the real order of natures grasped by the intellect.

Yes, the human mind begins as *tabula rasa* — nothing is in the intellect that was not first in the senses. But from repeated experience, the mind abstracts universals and discerns necessity. When we say, 'Every effect has a cause,' this is not habit (as Hume thought), nor an arbitrary mental law (as you claim). It is the intellect's recognition of what is necessarily true of being: that what comes into existence must be grounded in another.

In that sense, I will affirm the 'synthetic *a priori*' — but I reformulate it. *Our universal and necessary judgments are indeed synthetic (they extend knowledge) and a priori (they hold universally, not as statistical habits). But they are not imposed from the subject. They are discerned by abstraction from the natures of things.* The order of being impresses itself upon the intellect, and the intellect recognizes necessity as a reflection of what exists, not a construction of our faculties."

Kant: "When Jesus calms the storm, the category of causality no longer applies. Phenomena must follow law; miracles, if they exist, are noumenal and unknowable."

Aquinas: "Here the reformulated synthetic *a priori* shines. We know from experience and abstraction that storms have natural causes: □(Antecedents → Storm). That is a necessary truth abstracted from being, not imposed by the mind. But we also know the sea has a real potency for calm. Ordinarily this is actualized by natural causes, but it may be actualized directly by God. *The miracle does not destroy causality; it shows a higher cause directly actualizing a potency. The law holds, the miracle is intelligible, and necessity is real because grounded in being.*"

Socrates: "So let us apply this. Hume: storm follows from habit, not necessity. Kant: storm follows necessary law, but law is imposed by the mind. Thomas: storm follows natural potency, grounded in being, sustained by God. Which best accounts for both science and miracle?"

Aquinas: "***Miracle is possible because God can actualize potencies directly***. Chance is possible because finite causes intersect contingently. Necessity is possible because essences ground natural laws. All three — necessity, chance, miracle — are united under ***ipsum esse subsistens***."

Klima: "Consider the tornado. $\Box(A \rightarrow$ Tornado) given antecedents. $\Diamond$(Park A strike) $\wedge$ $\Diamond$(Park B strike). Chance is real, *per accidens*. God sustains both potencies. For Hume, this is only expectation. For Kant, the mind imposes order. But only classical metaphysics shows contingency and necessity grounded in being itself."

Dodds: "And in modern language: God sustains natural laws (statistical physics) and permits contingency (quantum chance, genetic drift). Providence includes both."

> **Socrates:** "So the weaknesses are plain:
>
> - Atomists deny teleology, leaving brute determinism or randomness.
> - Hume dissolves causality into custom, undermining science.
> - Kant saves necessity but imprisons it in the mind, making nature unknowable in
> - Kripke clarifies semantics but not ontology.
> - Plantinga explains providence but risks determinism by world-actualization.
> - Heisenberg shows real chance in physics but leaves no room for providence itself.

Aquinas: "The truth is this: *esse ipsum subsistens* sustains both laws and contingency. Chance is real in finite causes; miracles are real as direct divine acts; natural laws are real as potencies ordered to ends. Thus, storms, tornadoes, and miracles alike are intelligible without collapsing into determinism or skepticism."

Dodds: "And this framework even welcomes modern science: chance in quantum mechanics, contingency in evolution, order in natural law. Providence is richer for including both necessity and chance."

Socrates: "So the harmony is reached. Against Hume's skepticism, causality exists in the nature of things. Against Kant's transcendentalism, necessity is in being, and mind insofar as it is abstracted a necessary and universal. Against Atomist determinism, chance events to obtain. Against modern nominalism, universals are grounded in natures. Thus, Thomas, Klima, and Dodds, with Kripke's semantic clarity, secure philosophy, science, and metaphysics: divine providence sustains a world of law, chance, and miracle. In a word, *esse ipsum subsistens is Being and Necessity.*"

BIBLIOGRAPHY

Aquinas, Thomas. "De Ente et Essentia", ch. 4–5.

Aquinas, Thomas. *Quaestiones Disputatae de Potentia Dei*. In *Opera Omnia*, Leonine Edition, vol. 21. Rome: Commissio Leonina, 1918.

Aquinas, Thomas. *Quaestiones Disputatae de Veritate*. In *Opera Omnia*, Leonine Edition, vols. 22/1–2. Rome: Commissio Leonina, 1970–1972.

Aquinas, Thomas. *De Ente et Essentia*. In *Opera Omnia*, Leonine Edition, vol. 43. Rome: Commissio Leonina, 1976.

Aquinas, Thomas. *In I Sententiarum*. In *Opera Omnia*, Parma Edition, vols. 1–4. Parma: Fiaccadori, 1856–1858.

Aquinas, Thomas. "Summa Theologiae", I, q.3-11. Trans. Fathers of the English Dominican Province.

Aquinas, Thomas. "Summa Theologiae", I, q.14, a.5: "Whatever is in the intellect must be in it according to the mode of the intellect."

Ibid., I, q.15, a.1: "Ideas, according to Plato, are the exemplars existing in the divine mind."

Aquinas, Thomas. "Summa Theologiae." In "Sancti Thomae Aquinatis Opera Omnia," Leonine Edition, vols. 4–12. Rome, 1888–1906.

Aquinas, Thomas. "Summa Theologiae", I, q.2, a.3; I, q.5, a.4.

Aristotle, "Metaphysics", trans. W.D. Ross, Book XII. Oxford: Clarendon Press, 1928.

Aristotle. "Metaphysica." Edited by W. Jaeger. Oxford Classical Texts. Oxford: Clarendon Press, 1957.

Aristotle, Metaphysics, IV.3, 1005b19–34: "It is impossible for the same thing at the same time to belong and not to belong to the same thing in the same respect."

Aristotle. "Physics." Edited by W. D. Ross. Oxford: Clarendon Press, 1936.

Aristotle, "Physics'. Book II.

Aristotle. "Posterior Analytics." Edited by W. D. Ross. Oxford: Clarendon Press, 1949.

Augustine of Hippo. "*De Trinitate*. In *Corpus Christianorum Series Latina*," vols. 50–50A. Turnhout: Brepols, 1968.

Brower, Jeffrey E., and Susan Brower-Toland. "Aquinas on Mental Representation: Concepts and Intentionality." *Philosophical Review* 119, no. 2 (2010): 193–243.

Brower, Jeffrey E. "Making Sense of Divine Simplicity." *Faith and Philosophy* 25, no. 1 (2008): 3–30.

Craig, William Lane, and James D. Sinclair. "The Kalam Cosmological Argument." In *The Blackwell Companion to Natural Theology*, edited by William Lane Craig and J. P. Moreland, 101–201. Oxford: Wiley-Blackwell, 2009.

Davies, Brian. *The Thought of Thomas Aquinas*. Oxford: Clarendon Press, 1992.

Dowker, Fay. "Causal Sets and the Deep Structure of Spacetime", in "100 Years of Relativity", 2005.

Feser, Edward. "Five Proofs of the Existence of God." San Francisco: Ignatius Press, 2017.

Feser, Edward. "Scholastic Metaphysics: A Contemporary Introduction", Editiones Scholasticae, 2014.

Frege, Gottlob. "Über Sinn und Bedeutung." Zeitschrift für Philosophie und philosophische Kritik 100 (1892): 25–50.

French, Steven. "The Structure of the World," Oxford UP, 2014

Hughes, C. W. *Aquinas on Human Action: A Theory of Practice*. Washington, D.C.: Catholic University of America Press, 2012.

Hughes, G. E., and M. J. Cresswell. *A New Introduction to Modal Logic*. London: Routledge, 1996.

Hume, David. "An Enquiry Concerning Human Understanding." Edited by Tom L. Beauchamp. Oxford: Oxford University Press, 1999.

Kant, Immanuel. "Kritik der reinen Vernunft." Akademie-Ausgabe, Bd. III. Berlin: Reimer, 1900.

Kerr, Gaven. *Aquinas's Way to God: The Proof in De Ente et Essentia*. Oxford: Oxford University Press, 2015.

Klima, Gyula. *Aquinas' Theory of Modality*. New York: Fordham University Press, 1988.

Klima, Gyula. "The Medieval Problem of Universals." In *The Stanford Encyclopedia of Philosophy*, edited by Edward N. Zalta. Metaphysics Research Lab, Stanford University, 2022.

Klima, Gyula. "Aquinas on the Semantics of Modality." *The Modern Schoolman* 72, no. 1 (1994): 1–22.

Knuuttila, Simo. *Modalities in Medieval Philosophy*. London: Routledge, 1993.

Kripke, Saul A. "Naming and Necessity." Cambridge, MA: Harvard University Press, 1980.

Kripke, Saul A. "Semantical Considerations on Modal Logic." *Acta Philosophica Fennica* 16 (1963): 83–94.

Ladyman, James and Ross, Don. "Every Thing Must Go," Oxford UP, 2007.

Leftow, Brian. *God and Necessity*. Oxford: Oxford University Press, 2012.

Moreland, J. P. (2018). Scientism and Secularism. Crossway.

Nagel, T. (2012). Mind and Cosmos. Oxford University Press.

Oderberg, David. Real Essentialism, Routledge, 2007.

Owens, Joseph. "An Elementary Christian Metaphysics," Houston: Center for Thomistic Studies, 1985.

Pasnau, Robert. *Metaphysical Themes, 1274–1671*. Oxford: Oxford University Press, 2011.

Plantinga, A. (2000). Warranted Christian Belief. Oxford University Press.

Plantinga, Alvin. *The Nature of Necessity*. Oxford: Clarendon Press, 1974.

Quine, W. V. O. "Two Dogmas of Empiricism," Philosophical Review 60 (1951): 20–43.

Gilson, Etienne, "The Christian Philosophy of St. Thomas Aquinas", 1956.

Rovelli, Carlo. "Quantum Gravity", Cambridge UP, 2004.

Smith, Q. (1991). Theism, Atheism, and Big Bang Cosmology. Oxford University Press.

Stump, Eleonore. *Aquinas*. London: Routledge, 2003.

Swinburne, R. (2004). The Existence of God (2nd ed.). Oxford University Press.

Wippel, John F. "The Metaphysical Thought of Thomas Aquinas". Washington, D.C.: CUA Press, 2000.

APPENDIX A: MODAL BASE RULES, FRAME CONDITIONS, BRIDGE AXIOMS, AND BARCAN POLICY

Modalities
- $\square$ = metaphysical necessity (alethic, *de re* friendly)
- **K** = human epistemic knowledge (fallible agents)

1) Base rules (per modality)
- **Normality (K-axiom)**: $\square(A \rightarrow B) \rightarrow (\square A \rightarrow \square B)$; and $K(A \rightarrow B) \rightarrow (KA \rightarrow KB)$
- **Necessitation**: if $\vdash A$ then $\vdash \square A$; if $\vdash A$ then $\vdash KA$
- *(These give you normal modal logics for $\square$ and for K.)*

2) Frame conditions
- **Metaphysical $\square$: S5** (reflexive, transitive, symmetric). Classical metaphysical necessity tracks essence/possibility in virtue of natures; S5 is the standard strength used in classical metaphysics and is safe for *de re* work.
- **Epistemic K: S4** (T + 4). Knowledge is factive and positively introspective in idealized models for agents, but we **do not** assume negative introspection (so not S5 for human knowers).
  - o   T: $\square A \rightarrow A$; $K A \rightarrow A$ (factivity)
  - o   4: $\square A \rightarrow \square\square A$; $K A \rightarrow K K A$
  - o   **No 5 for K** ($\neg KA \rightarrow K\neg KA$ not assumed)

3) Bridge axioms (interaction principles)
To keep classical logic/metaphysical distinctions (contingent truths are truly contingent even if known by God; human knowledge needn't track necessity), we adopt **very conservative bridges**:

1. **Factivity of K**: Kp → p.
2. **Positive introspection**: Kp → KKp.
3. **Knowability of necessities (modest)**: □p → ◇Kp.
   Read: if p is necessary, it is in-principle knowable by rational agents.
4. **Rigid designation (names/k-inds)**: Rigid(a) and rigid sortals allow standard Leibniz substitution into □-contexts.
5. **Essentialist *de re* schema** (Aquinas-style): **Ess(F, x) → □(E!(x) → F(x)).**
   If F belongs to x per se/essentially, then if x exists, x is F in all metaphysically accessible worlds.
6. **No "omniscience ⇒ necessity"**: We **do not** accept Kp → □p nor □p → K□p as general bridges for human K. (They would collapse the contingent/necessary distinction that Aquinas insists on.)

Note:

(5) uses an existence predicate E! to keep essence distinct from existence: essences ground modality; existence is contingent.

4) Barcan policy (quantifiers + worlds)

Classical metaphysics grounds *possibilia* in the **divine ideas**. A clean formalization is:

- **Quantify over possibilia with a constant domain** (the set of possible individuals is fixed in mente Dei).
  Under this reading, adopt **both**:
  o **BF**: ∀x □φ → □ ∀x φ
  o **CBF**: □ ∀x φ → ∀x □φ

- **Existence-at-a-world** is expressed by **E!(x)** inside formulas (not by shrinking/expanding the domain). When reasoning **about actual existents at a world**, relativize with E!(x): unrestricted BF/CBF needn't be used (and often shouldn't be).

APPENDIX B: THE METAPHYSICS OF DIVINE IDEAS

Classical metaphysics ultimately grounds all modal possibility in divine intelligibility for ontological reasons. Some x cannot be if it is incoherent to be.

A thing is not possible if it entails a contradiction or lacks any coherent form that could be instantiated.

Here are three examples. Note a unicorn can be possible insofar as a horn and a horned animal is not incoherent. But a circular square is impossible to exist because it is incoherent.

1. A Square Circle

Definition: A being that is both a square (four right angles and equal sides) and a circle (no angles, continuous curve).

Why It's Not Possible:

- These attributes are mutually exclusive in form; their union implies a contradiction.

- Lacks intelligible form: There's no divine idea of a square circle because it's not intelligibly definable. You can't derive a nature or essence from it.

Conclusion: A square circle is not a possible being because it has no coherent intelligibility—it collapses into logical incoherence.

2. A Married Bachelor

Definition: A being that is both married and not married.

Why It's Not Possible:

- The concept negates itself. It's not just improbable—it's incoherent.

- Lacks intelligible essence: No form in divine intellect can contain mutually exclusive predicates simultaneously.

3. A Colorless Color

Definition: A being that is a color, yet has no color.

Why It's Not Possible:

- Being a color entails being perceptible via light wavelengths. "Colorless color" is not a privation like "invisible ink"—it is a direct contradiction.

- Lacks intelligible content: You cannot even imagine or define what this would be.

Key Principle from Aquinas (ST I, q.25, a.3):

"Whatever does not imply a contradiction is among those things in the Divine power. But what does imply a contradiction does not fall under the scope of Divine power."

That is, God cannot create a non-being or contradictory being, because it is not a thing at all. It's not "too hard for God"—it's not anything to begin with.

- A possible being = an essence that is coherent, intelligible, and logically non-contradictory.

- An impossible being = a concept that negates itself and thus lacks intelligibility and form. A rock too big for God to pick up has no being because it simply does not exist and therefore is logically incoherent.

Only that which has an internally coherent form can exist even potentially, and that coherence requires a divine intellect to conceive it.

Contradictory "objects" are not grounded in being, and thus are not possible. Such objects cannot exist because they are simply ontologically incoherent. It is not simply in the mind, it's ontologically impossible for such things to exist.

The natural order reflects divine intelligibility; irregular or chaotic alternatives are not metaphysically possible. A tornado is a chaotic random event, but they are regulated by natural laws that do not contradict one another. Ontologically, they appear random and chaotic, and indeed they are random events. But in fact they are governed by natural laws. Velocity, wind current, etc.

A tornado cannot both stop spinning and continue to spin at the same time and in the same respect because it is regulated by natural laws. These natural laws are regulated by the law of non-contradiction itself. And the law of non contradiction is itself regulated by a divine intellect. The divine intellect created the very laws that govern creation because creation itself was a rational coherent act.

In philosophy course it's asked, "Can God make a rock so heavy He cannot lift it? The answer is no, because such a thing is ontologically a contradiction.

An intelligent being must exist in all possible worlds to ground the intelligibility of any possible entity. Therefore, in every possible world, God must exist as the necessary, self-subsistent, intelligent cause of all that is possible and actual.

This is not a mere modal hypothesis but a metaphysical deduction from the nature of being and the structure of intelligibility itself.

1. What Are Divine Ideas?

In classical philosophy, divine ideas (*rationes*) are the exemplary causes of all things. They are the forms or archetypes in the Divine Intellect, by which God knows all that He can create. Not abstractions like Plato's Forms existing independently of God. Not purely logical constructs, but real archetypes in God's self-knowledge.

"God knows things not because they exist, but things exist because He knows them." (ST I, q.14, a.8)

So: the divine ideas are the intelligible content of all possible essences, as known by God in His act of knowing Himself.

2. Possibility: Grounded in Divine Ideas

Aquinas's view of possibility:

Something is possible if it is not contradictory and could be actualized by a sufficient cause. But what determines whether an essence is possible? Not just abstract logic, but:

- Whether it can be intelligibly conceived and grounded in being.
- And that intelligibility is eternally contained in the Divine Intellect.

Thus, the range of all real possibilities (possible essences) is grounded in God's self-knowledge. God knows every way His essence could be imitated finitely—and those are the possible natures of creatures.

Example:

- A unicorn is a possible being, because it is a combination of intelligible features.
- But that possibility is grounded in the divine idea of "horse" + "horned animal", which pre-exist as known possibilities in the divine mind.

So:

Nothing is possible unless it is first intelligible. Nothing is intelligible unless it is first known in the Divine Intellect.

This is why Aquinas says:

"God knows not only what is, but what can be." (ST I, q.14, a.5)

3. Necessity: Grounded in Divine Ideas

Now, there are different kinds of necessity in Thomistic metaphysics:

Type of Necessity	Example	Ground
Logical	2+2 = 4	Law of non-contradiction
Metaphysical	God exists	Grounded in essence
Physical	Fire burns	Grounded in nature

The key is that what is necessary in created things is so because:

- It derives from a nature or essence that is itself derived from the divine idea.

- Metaphysical necessity (e.g., God must exist) follows from the fact that God is *ipsum esse subsistens*.

So even necessity is intelligible because:

- It stems from what God knows as necessary by His nature (e.g., "being *qua* being cannot not-be").

- And any necessity in creatures is a reflection of God's creative intellect ordering things to ends.

4. Summary in Classical Terms

- Divine Ideas are the intelligible patterns in God's intellect.

- They are the ground of possibility, because what is possible must be first intelligible, and only God knows all that is intelligible.

- They are the ground of necessity, because God knows what must be by His own nature, and what must be by consequence of a given nature.

Without the divine intellect, there would be:

- No intelligibility,
- No meaningful essence,
- No measure of what could be,
- No explanation for what must be.